# Resonance

# Resonance: Electrical Engineering at the University of Rochester

Edwin Kinnen

MELIORA PRESS
An imprint of University of Rochester Press

First published by the Meliora Press
University of Rochester Press
668 Mount Hope Avenue, Rochester, NY 14620, USA
www.urpress.com

and at Boydell & Brewer, Ltd.
P.O. Box 9, Woodbridge, Suffolk 1P12 3DF
United Kingdom

ISBN 1-58046-142-5

**Library of Congress Cataloging-in-Publication Data**

Kinnen, Edwin, 1925–
Resonance: electrical engineering at the University of Rochester/
Edwin Kinnen.
p. cm.
Includes bibliographic references and index.
ISBN 1-58046-142-5 (Hardcover : alk. paper)
1. University of Rochester. Dept. of Electrical and Computer Engineering – History – 20th century. 2. Electric engineering – Study and teaching (Higher) – New York (State) Rochester – History – 20th century.
1. Title.
TK210.U56K56 2003
621.3′071′174789—dc21

2003010929

**British Library Cataloguing-in-Publication Data**
A catalogue record for this book is available from the British Library

Printed in the United States of America.
This publication is printed on acid-free paper.

This book is dedicated to my colleagues

# Contents


| | |
|---|---|
| List of Illustrations | ix |
| Foreword | xi |
| Preface | xiii |
| Acknowledgments | xv |
| 1. Introduction | 1 |
| 2. Before 1958 | 9 |
| 3. 1958 to 1970 | 21 |
| 4. The 1970s | 43 |
| 5. The 1980s | 59 |
| 6. The 1990s | 81 |
| 7. Closing Thoughts | 103 |
| Post Script | 105 |
| Appendixes | |
| A. IEEE Fellows | 107 |
| B. Ph.D. Degrees | 109 |
| C. M.S. Degrees | 125 |
| D. B.S. Degrees | 145 |
| Notes | 181 |
| Name Index | 185 |

# Illustrations


1. Rush Rhees Library xvi
2. Gavett Hall during and after construction 8
3. Rush Rhees 11
4. Power Engineering Laboratory, Gavett Hall, 1934 13
5. John Harrison Belknap 14
6. The Hopeman Engineering Building dedication 20
7. Daniel W. Healy 23
8. Preliminary plans for space usage in the Hopeman Engineering Building 27
9. Edwin L. Carstensen 29
10. David T. Blackstock 31
11.& 12. Faculty, graduate students and families at Webster Park, Rochester, Spring, 1964 32&33
13. Gerald H. Cohen and student 35
14. Robert C. Waag 37
15. Lloyd P. Hunter and students 38
16. Hugh G. Flynn 40
17. Paul Osborne and student with an ion implanter 42
18. Charles W. Merriam 47
19. Edward L. Titlebaum 50
20. Herbert B. Voelcker, Jr. 50
21. Joan Ewing 52
22. Dave Farden and student 55
23. Computer Studies Building 58
24. Sidney Shapiro 60
25. Alexander Albicki with Menghui Zheng and his wife, Chaolin 62
26. Thomas Y. Hsiang 63
27. Zeynep Çelik Butler and Donald Butler 63
28. Roman Sobolewski 64

29. Marc Feldman 64
30. Rochester Parapodium 70
31. Bruce W. Arden 71
32. Thomas B. Jones 73
33. Murat Tekalp 74
34. Victor Derefinko 74
35. Three figures from U.S. patents 80&82
36. Edwin Kinnen 82
37. Kevin J. Parker 83
38. Eby G. Friedman 84
39. Michael Wengler 85
40. Vassilios D. Tourassis 85
41. David Albonesi 86
42. Mark Bocko with graduate student 87
43. Alan Kadin with graduate student 88
44. Diane Dalecki 88
45. Jack Mottley preparing an endotracheal tube 90
46. Theophano Mitsa working with Professor Titlebaum on student project 91
47. Philippe Fauchet 92
48. High speed, low power HYPRES analog to digital converter 102
49. Zhe Zheng 105
50. Aleksandar Jovancevic and Zoran Mitrovski with new doctorate diplomas, 1999 108
51. Professor Parker with Vladimir Misic, 1999 124
52. B.S. class of 1997 144

# Foreword

The Department of Electrical Engineering (now Electrical and Computer Engineering) is nearing the half-century mark, and so this history by Professor Kinnen is both timely and important. The department has had a major impact on the careers and lives of almost two thousand alumni. And in the nearly fifty years of existence, it has launched many new ideas, concepts, and technologies that have changed the shape of the modern world.

One case in point is the development of "weak shock theory" by a young Professor Blackstock working on the third floor of the Hopeman Engineering Building in the 1960s. This development was a milestone event in nonlinear acoustics, which would have an impact on aerospace, sonar, and other fields. This was followed by work on nonlinear effects by other faculty members and students in subsequent decades. As a result, every high-end medical ultrasound scanner in the world in 2003 has a nonlinear focusing mechanism that produces better image quality than was ever before possible, particularly in hard-to-image patients. This is just one example of the accomplishments and impact of this remarkable department over the last four decades. Other examples are given in the text.

What attracted so many outstanding people to this department over the years? The students may have been attracted to one of the best student faculty ratios anywhere, along with the advantages of studying engineering in a university setting with links available to music, medicine, business, and unique "Take Five" opportunities. Faculty were attracted by the opportunity to teach small classes of great students, and by the department's philosophy of finding outstanding people and then supporting their highest professional ambitions.

Today the department leads the university in a number of key areas, in technology transfer, and in providing leadership for major research centers that cross departments and even universities. ECE students continue to pursue a remarkable range of career options. The innovative culture and values and promise of the department, documented in this history, along with the continuing enlistment of great students and faculty, bode well for the future and lead me to believe that the next fifty years will have even greater positive impact than the first.

*Kevin J. Parker*
Dean, School of Engineering and Applied Sciences

# Preface

Early on a January morning in 1967 I was descending from the heights of the Watchung Range to New Jersey's Route 22 on my way to the Newark Airport. I was to spend the day in Rochester, New York, looking over—and being looked over by—the Department of Electrical Engineering at the University of Rochester. After a short flight, Dan Healy, chairman of the department, met me at the airport and it seemed we had barely begun to talk when we arrived at the University. Dan outlined the way the day would go, who I would meet, etc. After the presentation on my research, there were opportunities for me to learn something about the department and its faculty.

I remember meeting Bill Streifer, Hugh Flynn, and David Blackstock. Ed Titlebaum, who has never lost his knack for making amusing comments, Lloyd Hunter, a solid-stater who had come from a distinguished career at IBM, and Ed Kinnen, as reserved then as he is now, were also among those I met. Altogether each member of the faculty was friendly, clearly knowledgeable and distinguished in his own area. In contrast to other universities I had visited, the attitude at Rochester seemed to be to add good people to the faculty and let them go on with what they wanted to do.

So why was Rochester different? What made it the kind of place that attracted and encouraged faculty who were confident in their ability to do good work and for whom the necessary obligations of being research entrepreneurs were seen as positives? And why were faculty from all over the university, including professional schools such as Medicine and Music, ready to give of their time and talent to answer one's questions and, if circumstances warranted, to join in cross-disciplinary research?

The university's small size seems to have led its leaders at all levels and over many years to appreciate the value and benefits of collaborative research. The intellectual excitement of taking just a short walk to be able to confer with colleagues in a variety of disciplines is hard to convey, but it is real nevertheless.

To preserve some record of Rochester's and EE's uniqueness is a worthy task, and Ed Kinnen has stepped forward to make it a reality. He has tried to identify as many people as he could who had some connection with EE from its second birth in 1958 to the end of the century. In keeping with his personal reticence, he has only occasionally included in his narrative the stories that lend character to the faculty he writes about. But it does not take much reading between the lines to realize where his heart is and why he put so much effort into producing this history.

To produce the record that he has, Ed has concentrated on faculty and on research. Yet he would agree that the students we faculty taught and involved in our research were a most important part of the culture of the department and the university. Working every day with bright young people was—and no doubt remains—both a challenge and a delight. To track the students—both undergraduate and graduate—whom we influenced at the beginning of their careers is an even more daunting task than the writing of the present history. Thus we can only hint at the subsequent success so many of our graduates experienced and must leave it for others perhaps to discover and to tell their stories.

Nor does this history attempt to do justice to the staff, so many of whom over the department's life span remained for so many years as stalwart supporters of our joint enterprise. But they are alive in our memories and they have added tremendously to the tone and tenor of the department.

All of us are indebted to Ed Kinnen for what he has accomplished. In compiling the information for this book, he has made it possible for those of us who were there to revisit the many memories. For those reading about the people and their exploits for the first time, this will be the first step toward future fond recollections.

So to all, good reading, and a tip of the hat to Ed Kinnen.

*Sidney Shapiro*

# Acknowledgments

Many thank yous to all the good folks at the University of Rochester who helped find records and the names of individuals no longer on campus. Maureen Muar, Administrator, Department of Electrical and Computer Engineering (1983–2001), and Nancy Martin, University Archivist and Rochester Collections Librarian, were especially helpful, as were Debra Neiner, Staff Accountant (1983–present), and Paul Osborne, Senior Technical Associate in the Department (1982–present). Thanks also to Rosemary Parker, Administrator, Office of the Dean, School of Engineering and Applied Sciences; Isabel Kaplan, Librarian, Carlson Library; Melissa Mead, Librarian, Rare Books and Special Collections; Francine Lyon, University Alumni Office; Sally Ann Hart, Office of Institutional Studies; Eileen Pullara, University Gifts Office; and Mary Bartholomew, Office of the Registrar. Pamela Clark, Public Relations Assistant for the Rochester Center for Biomedical Ultrasound, brought the images and text together and assisted with numerous edits as the copy evolved. Timothy Madigan, Editorial Director, University of Rochester Press, shepherded the project through production. Thank you to my colleagues who responded to innumerable email requests for information, and especially to Sidney Shapiro and Edwin Carstensen, as well as Charles Merriam, Kevin Parker, and Philippe Fauchet for reading and helping with draft material. Finally, the writing would not have started without initial encouragement from Dean Parker, who also found resources for the publication. Retirement made the writing possible and enjoyable.

**Figure 1.** Rush Rhees Library.

# 1: Introduction

University histories are often written about the people and the events that created the institution and made major contributions to its buildings and stature. Without detracting from the value of such broad-based histories, these volumes risk slighting the development and activities of specific departments and the individuals within those units who together in effect make the university. This chronicle looks at one department of the University of Rochester—the Department of Electrical Engineering—more recently named the Department of Electrical and Computer Engineering.

There are several arguments for writing a department history as distinct from a university history. A few of these arguments are intuitive. The original and many of the former members of a department faculty established the tone and direction of the unit and need to be remembered. Department chairs often reflect milestones as the academic focus and faculty interests change with time. Notable research results and publications might all be listed. Clearly students and short-term visitors would also be important parts of any department record. Students and visitors reflect the multiplying effect of a department as they disseminate the knowledge acquired during their time on campus.

There are other less intuitive arguments that support a narrative of the Department of Electrical Engineering at the University of Rochester. The accomplishments of the department as an entity are noteworthy for the relatively short time it has existed and for its small size, but also and in no small measure because of special circumstances.

The department existed within a small, private, university that focused primarily on liberal arts education and had a strong research orientation. The biological and physical science departments had commendable reputations. In an unusual layout, the buildings occupied by the university's School of Medicine and Dentistry and associated

Strong Memorial Hospital were adjacent to the main campus where the engineering faculties were located. The medical school was a research-oriented institution that had a long history of contributions to the science and practice of medicine. Also, the time the new department was organized coincided with the beginning of new interests in the potential applications of physical science to medicine. This was occurring at university medical institutions throughout the country.

It is also of more than incidental interest that the department came into existence after World War II. Inventions, theoretical studies, and innovations of electrical phenomena that had taken place in restricted military laboratories were being made public during the late 1940s and 1950s. As a result, curricula of many electrical engineering departments were beginning to change. Research opportunities were very different and exciting by comparison to those that existed before the war. Financial support was available for university research and there were many older and strongly motivated students ready to participate in the investigations. This department had the opportunity to begin new directions within electrical engineering as an entirely new faculty was being assembled. Departments with existing faculties at the end of the war were often burdened with senior members whose scholarly endeavors were in areas rapidly being superseded. For example, areas of instruction being dropped from crowded curricula were in electrical power and electrical machinery, vacuum tube electronics, and intricate linear circuit theory. Emerging areas of instruction deemed to have greater relevance were in systems and communication, and soon after in solid-state electronics and digital computers.

A final argument in support of a historical record of this department follows from the liberal arts character of the university's undergraduate programs and the fact that most students were living on campus. This had the effect of limiting the number of students enrolled in electrical engineering and consequently the size of undergraduate classes. The number of technical courses that undergraduates could take, along with required liberal arts courses, was also limited. As a consequence, there was time available for each faculty member to develop research interests in newly emerging areas of electrical engineering and in associated graduate courses. The early approval of Ph.D. programs in the recently established College of Engineering and Applied Science further encouraged the faculty to engage students in their research.

This history records both the development of research activities of the department faculty and to a lesser extent the evolution of courses in electrical engineering. It also lists the names of the literally thousands of individuals associated with the department over the years. This is a story about the influence that individual electrical engineers have had on colleagues both in the department and throughout the university. Most dramatic in this sense is the multiplying effect that biomedical engineering has had throughout the university. While biomedical engineering, per se, isn't within the subject areas that usually prescribe electrical engineering, the impetus for biomedical engineering to develop to the extent that it has at the University of Rochester came largely from within electrical engineering. As the department's activities in this interdisciplinary field were so extensive and involved so many individuals, biomedical engineering has been given more attention in the narrative than other areas of research.

This history is also a chronicle about the members of the department over more than four decades after World War II while they were initiating, developing, and adapting their research to keep abreast of the active areas of investigation in this country and beyond. It records resources that were obtained and maintained as the university and outside agencies oscillated between periods of expansion and increased financial support, and periods of contraction and financial constraint.

The focus of the later chapters is clearly on research. In broad terms, research wherever it occurs is the lode of electrical engineering. This is often more visible when it occurs in the industrial world; the invention of the transistor is a prime example. Even though teaching is usually seen as the bread and butter of the faculty, participating in research can motivate and stimulate individuals in their roles as teachers and mentors. It is often said that the most valuable intangible a professor can give a class is a sense of his or her personal enthusiasm for the subject at hand. When this derives from his or her research, such enthusiasm is truly contagious. And this enthusiasm is particularly valuable when it is conveyed along with the many abstract and conceptual topics of electrical engineering.

From another perspective, a research-oriented department such as this one is a place where the emphasis is almost entirely on individual effort. Therefore, it is reasonable to view this department as a collection of entrepreneurs. That the department succeeded as a whole bespeaks of the role of individuals in identifying viable problems for their research, to obtain support to develop the area, and to attract graduate students to participate in the work. Universities, in contrast to most industrial organizations, are places where new endeavors are not all expected to succeed. Professor Sidney Shapiro gained a bit of a reputation at one point for stating openly that a university is indeed a place where one is permitted to fail—and by implication to start again. And this applies not only to students but members of the faculty as well. Perforce, a university can also fail in the sense of a new undertaking and begin again, as is discussed in chapter 2.

The writer has been a part of the department for more than thirty years. This has provided ample time to observe the exaggerated effects that individuals and personalities have had on the evolution of the department—and further, to realize more fully the intrinsic worth of cooperation among individuals on the faculty, when it does occur.

The process of gathering names of individuals who have played a part in the growth the department began in 1990 when the writer was the chair of the department. It was clear at the time that a record of department faculty, students, staff, and visitors would be of more than casual interest sooner or later. The project remained incomplete until Dean Kevin Parker provided the incentive to get the project started again. The narrative grew as an extension from the search for names.

Unfortunately, many documents of the 1950s and 1960s no longer exist. Obtaining accurate lists of graduates and visitors turned out to be a challenge, notwithstanding the many computer-generated files existing in different administrative offices. In the end, about 10 percent of the names of B.S. and M.S. recipients have been included on evidence not necessarily unreliable but nevertheless on records that could not be confirmed in the Office of the Registrar.[1] Accounts of visiting scientists and professors from outside the country were not always consistent with respect to spelling, the times of appointment, or identification of the home institutions. The title of an appointment often changed during the stay. Dates given for awarding Ph.D. degrees can vary by a year as a result of differing dates that appear on the theses, in the department records, and in the programs printed for graduation ceremonies. Timely information about members of the faculty and courses of instruction contained in the university college bulletins is not always accurate due to the time lapse between the submission of the material and the printing of the bulletin. Bulletin information was used when other sources were not found.

Faculty and students of the department have had many interactions with colleagues in other parts of this and other universities. Names of these colleagues are recorded, sometimes at the expense of readability.

A few historical events of the university are included in this presentation. The reader is referred elsewhere for history at the university level; see *Rochester, the Making of a University*, by Jesse L. Rosenberger; *A History of the University of Rochester, 1850–1962*, by Arthur J. May; and also *75 Years of Chemical Engineering at the University of Rochester 1915–1990*, by John Friedly. The last is of interest for both history of the university and history of that department.[2]

Something needs to be said to defend the premise that the department is small relative to its impact. The premise derives from a comparison of the numbers of faculty members in departments of electrical engineering and the numbers of Ph.D. degrees awarded at a selected set of universities, as shown in the table. The departments selected for comparison are those that members of this department have at various times identified for personal cooperation or competition, or have been used for comparing student admissions and the placement of graduates. The table includes data for years early in the department's existence and toward the middle and the end of the time covered by the narrative.

**Comparison of Faculty Size in Departments of Electrical Engineering***

| | Number of Faculty and Number of Doctorates Awarded | | | | | |
|---|---|---|---|---|---|---|
| | 1965–66 | | 1980–81 | | 1996–97 | |
| | faculty | degrees | faculty | degrees | faculty | degrees |
| Carnegie Mellon Univ. | 28 | 20 | 24 | 10 | 34 | 29 |
| Cornell Univ. | 46 | 13 | 38 | 14 | 37 | 24 |
| Johns Hopkins Univ. | 21 | 8 | 15 | 4 | 15 | 8 |
| Princeton Univ. | 21 | 4 | 22 | 8 | 26 | 20 |
| Rensselaer Polytech. Inst. | 5 | – | 39 | 6 | 35 | 13 |
| Stanford Univ. | 56 | 35 | 55 | 49 | 42 | 78 |
| *Univ. Rochester* | *13* | *6* | *(12)*† | *(1)* | *15* | *(14)* |

*The 1965–66 data are from the *Annual Directory of the American Society for Engineering Education for 1967*, vol. 57, March 1967, p. 545. The 1980–81 data are from *Engineering Education*, vol. 72, March 1982, p. 427, and the 1996–97 data from *Engineering Trends*, www.engtrends.com.

†The numbers in parentheses were by count from department records.

With few exceptions, this history doesn't begin to describe the activities of the department's approximately fifteen hundred students. It doesn't mention their participation in faculty research projects or the prizes and awards they won. Nor does it mention the graduate and professional schools they entered, the prominent positions many attained in industry or at universities, or the companies they started. The narrative doesn't recall most of the staff who over the years contributed to the making, repairing, moving, typing, accounting, computing, and administrating that goes on daily in the life of the institution. This is truly unfortunate. However, records from earlier years have disappeared. The decision was made to omit these parts of the story rather than include the names and contributions of just a few or only those most recently a part of the department.

For reasons both practical and personal, this history does not go beyond the year 1999. While a cutoff date was necessary and this one was a somewhat arbitrary choice, the transition from 1999 to 2000 can be seen as a mark for fundamental changes occurring in and around higher education. Changing sources of income and support are directly and indirectly affecting literally every aspect of the university. Engineering education is responding to the ubiquitous and overpowering effects of the digital world. New graduates of electrical engineering in the twenty-first century can be expected to have less in common with their predecessors. Areas of academic research will reflect few of the concerns that initiated investigations during the 1980s and 1990s. The twenty-first century may well require a new perspective to record and evaluate the happenings of a contemporary department of electrical engineering, a task for someone else.

## A Note on Sources

Some of the information in the following chapters is duly referenced. Otherwise material came from the reports of the Dean of the College of Engineering and Applied Science from the years 1960–63, 1965–71, and 1988–93; annual reports from 1982–83; the department's alumni newsletter, the *EE Network*; college catalogues; programs printed for graduation ceremonies; annual reports of the Rochester Center for Biomedical Ultrasound, department files, and personal conversations.

The photographs are from the University Public Relations department, the archives of the Rare Books and Special Collections of the Rush Rhees Library, department files, and from private collections.

"Strange how much you've got to know
Before you know how little you know."
*Anonymous*

**Figure 2.1.** Crews working on the new engineering building, September 17, 1929.

**Figure 2.2.** Work continued on October 15, 1929.

**Figure 2.3.** This building would be named Gavett Hall.

# 2: Before 1958

It is easy to be wise after the event.
*English proverb*

The Department of Electrical Engineering at the University of Rochester was established in 1958 by the board of trustees. Prior to 1958, members of the faculty of the university had been offering courses in electrical science and electrical engineering for many years. A Department of Electrical Engineering had also existed for a short period in the late 1940s. This chapter reviews early events preceding the creation of the department in 1958.

Interestingly, the origins of the University of Rochester in 1850 overlap, in time, an event that can be viewed as the origin of electrical engineering as a distinct discipline. By the 1850s, Michael Faraday in London had already performed his experiments on electromagnetism, Signor Salvatore dal Negro of Padua had constructed the first oscillating motor, and M. Hypolite Pixii of Paris had created the first electric generator. Items such as the motor and the generator were part of the 1851 Exhibition in the Crystal Palace in London, along with descriptions of the channel cable laid down between England and France and early attempts at providing illumination using electrical current. Consequently, the Exhibition has been seen by some as providing the first occasion for electrical engineering's public debut as a recognized profession.[1] (It can also be argued that the profession, per se, was established in 1884, coinciding with the appearance of the International Electrical Exhibition in Philadelphia.[2])

Also within ten years of the founding of the University of Rochester, many individuals were born who would make major contributions to the field of applied electromagnetism: Thomas Alva

Edison (b. 1847), Alexander Graham Bell (b. 1847), John Hopkinson (b. 1849), Oliver Heaviside (b. 1850), Silvanus P. Thompson (b. 1851), Elihu Thomson (b. 1853), Heinrich Hertz (b. 1857), and Nikola Tesla (b. 1857). Briefly, Edison is known most often for the light bulb and the phonograph, Bell for the telephone, Hopkinson for dynamo machines, Heaviside for the ionosphere, Thompson for alternating current systems, Thomson for the electric energy meter, Hertz for electromagnetic radiation, and Tesla for the induction motor. All together, this was an auspicious time.

Catalogues of the University of Rochester, from 1850/51 to 1893/94, contain no mention of course work that included electrical phenomena. The catalogue for the next academic year, 1895, shows a course in advanced electricity "intended for students who are making a specialty of physics or of electrical engineering."[3] A few years later, in 1899/1900, three courses in electrical science were mentioned. This is the year women were first admitted to the university and Henry F. Burton was acting president. Henry E. Lawrence, Harris Professor of Physics, offered these courses: Elementary Electricity, given during the fall quarter, Dynamos and Motors during the winter quarter, and Alternating Current Experiments during the spring quarter. They were stated later to be part of a pretechnical studies program intended for students who in their third year would enter "technical schools of the highest grade and so win a college and a technical degree in six years."[4] From then on, topics in electrical science would appear in the university's curriculum.

The third president of the university, Benjamin Rush Rhees, referring to the then new Eastman Laboratories for physics and biology in 1906, stated that "Work upon this new building has progressed slowly.... Professor Lawrence has been conducting the work of his class in Dynamo and Motor in the new Dynamo Laboratory during the spring term in as much as the class was so large that it was impossible to accommodate it in the old laboratory in Anderson Hall."[5] Later in this same report, purchase orders can be seen for equipment for the Dynamo Laboratory, including orders for seven motors, five generators, a rotary converter, and transformers. One of the motors, a 230-volt, DC, 7.5 HP unit, was ordered from the Rochester Electric Motor Company.

Five years later, instruction in electrical science would be given in the Carnegie Laboratory, another new building on the Prince Street campus in Rochester. The existence of the Carnegie Laboratory is attributed to President Rhees. According to Jesse L. Rosenberger in *Rochester: The Making of a University*, Andrew Carnegie made an offer to the university in 1905 "to give $100,000 for the erection and equipment of a building for applied science, on condition that the university should raise another $100,000, to be added to its endowment." This offer, writes Rosenberger, "meant an invitation to the university to branch out into a new field and to give, in addition to the course of instruction being offered, full training for students in mechanical and electrical engineering, or the scientific applications of power in modern industry."[6] By 1908, President Rhees had raised the second $100,000 and plans were started for the Carnegie Laboratory, the sixth building to be constructed on the Prince Street Campus.

During a leave from the university in 1908–09, President Rhees had a particular interest in visiting Georg August Universitat in the German town of Göettingen, "because there more than at any other German university, serious attention is being given to the intimate connection which should exist between applied science and the pursuit of the more familiar lines of liberal culture."[7]

When President Rhees returned from his sabbatical in Germany, he wrote, "My interest here is rather in reporting what is to me a most significant educational development, in the recognition of (1) the contributions which applied science has to offer to the study of science for its own sake, and (2) the advantage which such study of pure science may find in the methods presented by practical engineers and leaders in scientific industry. This educational development cannot stop with a modification of our treatment of physical science. It must influence our whole attitude in education, banishing the fear that culture will lose its fineness and its power if it brings itself to the test of practical efficiency and subjects its processes to the criticism of the practical experience of later active life."[8]

**Figure 3.** Rush Rhees.

Engineering at the University of Rochester owes much to President Rhees for the early physical manifestations of his efforts. But in a more lasting sense, President Rhees encouraged a liberal attitude toward applied science in an academic community that had recognized only pure science, noting "the attitude of universities toward the technical schools is still one of undisguised condescension and often dislike."[9]

Millard Clayton Ernsberger came from Cornell University in 1909 as Professor of Mechanical Engineering and Head of the Applied Science Department in the College of Arts and Science. He had received an undergraduate degree in the classics at Rochester in 1888 before going to Cornell and gave impetus to the philosophy of technical training in a general liberal arts culture. Professor Ernsberger was instrumental in the planning and early use of the Carnegie Laboratory, and also in establishing a six-year degree program in engineering. This program was quickly reorganized into a four-year degree program for a B.S. degree major in mechanical engineering. The university graduation program for 1914 lists three individuals receiving this degree: Howard Elston Bacon and Roy Hulme Hendrickson, Class of 1913, and Frederick James Converse. During the following years, Professor Ernsberger campaigned for another new building, one just for engineering. When his efforts failed to generate enough interest, he resigned in 1921 and returned to Cornell.

The 1909–1910 university catalogue notes, "there are... no college dorms, boarding and rooms can be obtained... from $4.50–6.00 a week,... [T]uition [is] $25 a term, [and the] fee for incidental expenses—such as janitor service, heat, light, repairs and use of the gymnasium—is $7 a term. Additional fees, to cover the cost of fuel, power and materials consumed in their work, are charged to students who take work in the chemical, biological, geological and physical laboratory." An average year's expenses were estimated to be $337 including room and board.[10]

When Professor Ernsberger resigned in 1921, Joseph W. Gavett, Jr. (M.E. Cornell, 1911) came from Cornell University to be Professor of Mechanical Engineering and Chairman of the Department of Engineering. Nine years later, the College for Men moved from the original Prince Street Campus to the River Campus. (The College for Women remained at Prince Street.) The new campus included an engineering building separate from the main quadrangle. Designed for the Departments of Mechanical Engineering and Chemical Engineering, the building would later be named Gavett Hall.[11]

It appears that in 1931 James Albert Wood, Jr., Instructor in Mechanical Engineering, took over the courses that Professor Lawrence

**Figure 4.** Power Engineering Laboratory, Gavett Hall, 1934.

had been teaching in the physics department. Wilbur Reed LePage, however, was listed as the instructor of these courses during the 1933–34 academic year. The following semesters, LePage introduced two new courses, Industrial Applications of Electricity and Principles of Electrical Communication, that included material on vacuum tubes. (Some years later, LePage may have made a more significant contribution to electrical engineering on campus; see chapter 3.) Charles Holcomb Dawson, appointed an Instructor in Electrical Engineering in 1938, replaced LePage in these courses. Gordon James Watt (B.S.) joined the faculty as an Instructor in Electrical Engineering in 1944 while Dawson was on leave in military service. Mr. Watt offered five courses in electrical engineering, again principally for mechanical engineering students.

In 1944, along with instruction given by the Department of Mechanical Engineering, the university offered special off-hour courses sponsored by the U.S. Office of Education. These were Science and Management War Training courses for the Rochester community, said to be at the college level but without academic credit. They included Elements of Radio Communication, Part I and Part II; Fundamentals of Electric Waves; Elementary Electronics; and Broadcast Station

Engineering. Instructors for the Radio Communication courses included Robert Bechtold from the General Railway Signal Company and Guy O. Crandall from the Civil Aviation Authority (CAA) Range Station at the Rochester airport. Warren Wheeler and Alfred Balling from WHAM Radio (Rochester) were the instructors for the electronics course, and Kenneth Gardner from WHAM and Bernard O'Brien from WHEC Radio (Rochester) gave the lectures for Broadcast Station Engineering. The Navy V-12 program on campus during this same period would evolve into the Naval Reserve Officer Training Corps (NROTC), which exists to the present.

Professor Gavett died in 1942 and a search was started for a replacement to head the Department of Engineering. An offer was made subsequently to Colonel John Harrison Belknap to chair the newly designated Division of Engineering in the College of Arts and Science. At the time—March 1945—Col. Belknap was still in military service.

Professor Belknap did accept the offer and arrived on campus during the summer of 1945 after discharge from the military. The Department of Electrical Engineering was formed quickly as there was an immediate need to provide instruction in new areas of electrical engineering that had emerged during the war effort, particularly in electronics and communication. This third engineering department in the division was chaired by Professor Belknap and in addition consisted of Mr. Dawson and Mr. Warren Ray Wheeler as instructors.* The university's catalogue stated that tuition for that year was "$250 per term with annual costs, including room and board, estimated to be $1350 for men and $1400 for women."

**Figure 5.** John Harrison Belknap.

However, this new department was short lived. Veterans of World War II were returning to universities everywhere during the fall of 1945

* Optics at this time was a separate department in the College of Arts and Science.

under Public Law 346, the Serviceman's Readjustment Act, better known as the G.I. Bill. At this time, engineering students at the university increased from 181 the year prior to 414, including 41 freshmen entering the new department. The combination of the increased engineering enrollment and the remaining V-12 future naval officers overburdened both the faculty and facilities. Coincident with this, the total operating costs for the Division of Engineering were predicted to increase from $55,000 in 1944–45 to $135,000 in 1948–49.[12] The College of Arts and Science decided that it could not shoulder the increased costs, including those for establishing and maintaining the laboratories that were requested by the new department. In May 1948, the university's board of trustees decided that no further electrical engineering student majors could be admitted and that the department would be eliminated as of June 1950. Professor Belknap promptly resigned from the university and became dean of the graduate section of the U.S. Air Force Institute of Technology at the Wright-Patterson Air Force Base in Dayton, Ohio. Ironically, the solid-state transistor, which was destined to have a profound effect on electrical engineering, had been invented at the Bell Laboratories in 1947. In the decades following, most familiar products operating with relays or vacuum tubes would be redesigned using small, low-power transistors. Some, such as radios, television, and computers, would eventually undergo extreme miniaturization using solid-state integrated circuits.

The action of the board received considerable publicity at the time and in retrospect may be difficult to understand as a laudable decision. The trustees had been discussing an expansion of engineering at the university and possible attendant costs since the early 1940s. University Council minutes of June 18, 1945, note that advice given by national leaders in engineering, both in academia and in industry, included the advisability of adding a four-year course in electrical engineering. Alan Valentine, president of the university since 1935, had consulted with twenty leaders of local industry and clearly recognized the postwar needs of a large number of veterans and civilians for an engineering education.

So what went wrong? Who was John Harrison Belknap and how did this reversal happen? Professor Belknap was appointed in March 1945 after a two-year nationwide search. His credentials were noteworthy. Born in 1892, he received a B.S. degree with a major in electrical engineering from Oregon State College in 1912 and then taught there. He joined the Westinghouse Corporation in Pittsburgh in 1923 after military duty in World War I. His career at Westinghouse was varied; at its end he was manager of the corporation's engineering training program. He was active in engineering education societies at the national level, and received an Honorary Doctorate of Engineering degree from his alma mater in 1940. He re-entered military service, in the U.S. Air Corps, in February 1942. In August 1945, following the armistice, then Lt. Col. Belknap was assigned to duty with the U.S. Group Control Council in Germany as deputy chief of the Electrical and Communications Branch.[13] There were letters from President Valentine to Col. Belknap during the last days of his military obligations stating that the university was looking forward to his being on campus for the fall term. Other letters were sent to Washington personalities asking that Col. Belknap's discharge be expedited if possible.

When Dr. Belknap arrived on campus, he was appointed Yates Professor of Engineering, an endowed professorship that today is still held by an outstanding member of the engineering faculty. Notices went out to local business leaders about his arrival and he gave numerous lectures to local engineering societies on his experiences in postwar Germany. Clearly the university had worked hard to find someone of his stature and background, was pleased to attract him, had made plans for postwar growth in engineering, and was quick to provide publicity to this effect. As the successor to Professor Gavett, he was to start a Department of Electrical Engineering and also to head the Division of Engineering and lead the engineering effort of the university in the postwar period.

President Valentine had recognized the postwar needs of a large number of veterans and civilians for engineering education and understood that a degree program in electrical engineering was particularly important. It was also understood that this would require additional faculty and more laboratory equipment both for electrical and mechanical engineering. In a September 1946 letter to Raymond L. Thompson, University Treasurer, about changes in the engineering budget proposed by Professor Belknap, J. Edmund Hoffmeister, then Dean of the College of Arts and Science, stated, "I feel that Belknap is reasonable about most of his requests.... If all new heads of departments were as reasonable, we would have fewer headaches." However, in view of Professor Belknap's plans for engineering and with student numbers increasing rapidly, Dean Hoffmeister suddenly became alarmed and administrative support for Professor Belknap faded. While President Valentine's correspondence with Professor Belknap during this period showed understanding, patience, and attempts at accommodations, a blow-up finally occurred. Letters from the president to Professor Belknap then included terse statements such as "present costs and added cost predicted by you are heavier than the University can bear" and "you get the money yourself." One can suspect a volatile personality finally clashing with another honed by years of military service.

The conclusion seems clear; realistic budgets were not discussed early on when plans for the new department were developing, or during meetings with local business leaders, or during interviews with Col. Belknap. In the end, the university needed significant extra financial support from local industries that simply did not materialize. Arthur May, in the manuscript version of his book, *A History of the University of Rochester, 1850–1962*, adds another slant to this unfortunate incident, namely that too many male students were enrolling in engineering, thereby threatening the atmosphere of the liberal arts college. (An excerpt from the manuscript appears at the end of this chapter.)

Whatever the politics of the moment, Professors Belknap, Assistant Professor Dawson, and Instructor Wheeler offered twenty-three electrical engineering courses in 1947, including courses in electronics, industrial control, communications, electrical machinery, transmission and distribution, illumination, and insulation. Then in 1948, John Baird (Sc.D.) and Arlie E. Paige (M.S.) joined Dawson as assistant professors, and a year later the name Robert Earle Vosteen (M.E.E.) also appeared as an instructor along with Mr. Wheeler. After Professor Belknap left (followed by Professors Paige and Baird), the twenty-three courses were reduced to seventeen and in 1950 to seven as a smaller faculty responded to student interests as well as they could with limited resources. The courses in illumination and insulation were dropped, as were those in electrical machinery even though one in servomechanisms was added the next year. While Robert Q. Pollard (B.S.) joined Wheeler and Vosteen as instructors in 1950, only Professor Dawson and Mr. Pollard remained for the following two years. The university catalogues continue to show Professor Dawson, but with Roy Clifford Johnson (B.S. Rochester) in 1953 and 1954, and then with Barry Carl Dutcher (M.S. Rochester) during the next three years as the only individuals providing instruction in electrical engineering. Both Mr. Johnson and Mr. Dutcher, however, had formal appointments as Instructor in Mechanical Engineering. But courses in 1957 were again said to be offered in circuit analysis, electrical machinery, transients in linear systems, electronics, filters, fields and waves, industrial electronics, and feedback control, plus two courses each for mechanical and chemical engineering students.

This first effort to establish a Department of Electrical Engineering, nevertheless, did produce some graduates. The 1949 graduating class of the College for Men included seventeen individuals who received a B.S. degree with electrical engineering as a major.

| | |
|---|---|
| Zygmund J. Bara | Harry R. Nichles |
| Clement O. Bossert | Joseph Phillips |
| Ralph J. Brown | Robert Q. Pollard |
| Philip J. Buchiere | Elliott I. Pollock |
| Donald P. Dise | Walter J. Randolph |
| Thomas E. Doughty | Bernard J. Schnacky |
| Robert J. Hoefer | Cecil E. Scott |
| Robert P. Kennedy, Jr. | Grosvenor S. Wich |
| Ronald A. Miller | |

Another sixteen individuals graduated with a B.S. degree with a major in electrical engineering in 1950 at the one hundredth Commencement of the university.

| | |
|---|---|
| John V. Adkin | Robert C. Iseman |
| Wilbur E. Ault | Richard F. Kaiser |
| Irwin S. Booth, Jr. | Nicholas Lazar |
| Edward S. Brown | William G. Nyhof |
| Ingvar E. Eliasson | Arthur R. Principe |
| Albert C. Giesselman | Erick N. Swenson |
| Lewis M. Goodrich | Robert B. Taylor |
| William G. Graeper | Marvin Trott |

John Frank, a holdover from the previous class, graduated the following year with his B.S. degree with a major in electrical engineering. No degrees with a major in electrical engineering were granted during the following six years, 1952 to 1957.

The Division of Engineering continued as a somewhat autonomous part of the College of Arts and Science until 1958. Lewis D. Conta (Ph.D. Cornell, 1942), Professor of Mechanical Engineering, was chair of the division after Professor Belknap resigned and would also serve as the acting dean of a new College of Engineering until the new dean arrived. The curriculum during these years emphasized theory and fundamental science concepts, with students entering the college after two years in the College of Arts and Science. Interestingly, Dean Conta is remembered for encouraging all engineering faculty during the late 1950s to develop digital computer skills.

---

The following passage, obtained from the University of Rochester archives, appears in the manuscript copy of Arthur May's "A History of the University of Rochester 1850–1962," chapter 32, pp. 21–22:

> An ambitious plan to expand the division of engineering by adding instruction in electrical engineering had a chequered career. Started in 1945, electrical engineering enrolled a large number of students, especially military veterans.... J. Harrison Belknap... arrived in Rochester in the spring of 1946. [In fact Professor Belknap arrived during the summer of 1945.] It was soon evident—painfully so—that the costs of instruction and equipment in electrical engineering far outran advance calculations; the income from an endowment of around one million dollars, it was discovered, would be required. Unless interested Rochester industries were willing to underwrite the program, wisdom dictated that it should be discontinued; a good share of deliberations on the subject were not put on paper, though it is clear that in some University circles it was felt, quite apart from the financial aspects, that a disproportionately high percentage of the men undergraduates enrolled in engineering. Consequently, the University trustees voted to eliminate electrical engineering as of 1950. That decision, interpreted as a breach of faith, evoked a storm of resentment and unfavorable publicity, highly critical of the University in general and President Valentine particularly. To the protesting students Valentine seemed "the typical army officer"—issuing orders from the top—doing little to achieve the camaraderie and loyalty of his men; they flatly doubted whether the financial angle was the basic cause for an arbitrary reversal of policy.... In high dudgeon, Belknap retired from the University stage.

The published book, *A History of the University of Rochester 1850–1962* (Princeton, N.J.: Princeton University Press, 1977), differs significantly in several places from the original manuscript due to editing and abridging by Lawrence Eliot Klein. Except for two paragraphs that mention the department or College of Engineering, nothing is said in the published version about Mr. Belknap or the brief existence of the Department of Electrical Engineering. The edited version, however, does note that President Valentine had given warnings about possible financial difficulties in postwar years and described struggles the university had in the late 1940s while trying to respond to the physical and financial problems caused by the very large increase in enrollments.

**Figure 6.1.** The dedication of the Hopeman Engineering Building. *Left to right*: Albert A. Hopeman, Jr., Arendt Hopeman, Dean John W. Graham, Jr., President W. Allen Wallis, the Reverend Carlton C. Allen of Twelve Corners Presbyterian Church, who gave the invocation, and Joseph C. Wilson, class of 1931, chairman of the university's board of trustees.

**Figure 6.2.** Hopeman Engineering Building with cherry trees in blossom.

**Figure 6.3.** Friends, faculty, staff, and students attending the dedication ceremony.

# 3: 1958 to 1970

> "We must always have old memories and young hopes."
> *Houssage*

In 1958, the year the present department was created, The College of Engineering was established as a separate college of the university. Cornelius W. de Kiewiet (Ph.D. London, 1927) had been President since 1951. Prior to this, he was the provost and acting president of Cornell University as well as an active historian, principally in African history. During his ten years at the University of Rochester, the College for Men and the College for Women were united on the River Campus and three new colleges were established: Education, Business Administration, and Engineering.

John W. Graham, Jr. (D.Sc. Carnegie Mellon, 1950) came from the Department of Civil Engineering of Cooper Union to be the first dean of the new College of Engineering. At that time, Professor Conta was appointed the Associate Dean for Graduate Studies. Dean Graham's goals, as stated in his *Objectives for the College*, were

> to embrace the achievement of balanced programs in undergraduate education, graduate education, research, and service to community and industry—where each not only is significant in its own right, but contributes to the vitality of the others as well. The College seeks excellence in the four principal fields of engineering and applied science in which it offers degrees: chemical engineering, electrical engineering, mechanical engineering, and optics. The students are pointed specifically toward those functional careers that are uniquely engineering: high-level design, development, research, and teaching. In all of the activities of the College primary emphasis is placed on the quality of the work done, rather than on the number of students trained or the number of research

> papers published. The success of the College in achieving excellence is dependent more than anything else upon the caliber and productivity of this faculty.[1]

The original goals of the College of Engineering are noteworthy as they set a framework that still exists. In effect, the goals directed the faculty to a science-oriented undergraduate degree program that would prepare students for graduate study as well as professional employment over a lifetime when technology would be changing continuously. In practice, the undergraduate curriculum from the very beginning would consist of four courses per semester. This was a significant departure from the five- or six-course load typically required in engineering programs at most other universities. Limiting courses to four per semester had the effect of forcing instructors to carefully re-examine the contents of their courses as the technology evolved, instead of simply proposing new courses. Also with four courses per semester, over half of the thirty-two courses required for the B.S. degree would be taken in the College of Arts and Science. As a consequence, teaching assignments would be relatively low and scholarly activity would be expected from each member of the faculty.

The beginnings of the Department of Electrical Engineering occurred fortuitously in an expanding urban region that contained some large high-technology industries, particularly in the fields of optics and image reproduction, namely the Eastman Kodak Company and the Haloid Xerox Corporation. A university news release stated: "Major factors in the decision to establish a Department of Electrical Engineering were the educational advantages it affords, the strong interest of local industries in the establishment of such a curriculum at the University of Rochester, the increasing importance of that field of engineering in the new age of automation and electronics, and the outlook for higher enrollment in engineering at the University over the next 10 years or so."[2] Several individuals from these industries would later join the department.

Daniel Ward Healy (Ph.D. Harvard, 1951) was recruited from the University of Syracuse in September 1958 to be the chair of the new Department of Electrical Engineering. Interestingly, Wilbur LePage, who had been teaching electrical engineering subjects at the University of Rochester, 1933–38, joined the Electrical Engineering faculty at Syracuse University after receiving a Ph.D. degree from Cornell University in 1941. Professor LePage's tenure as chairman of that department, 1956–74, overlapped the years that Professor Healy was also at Syracuse, 1951–58. It is likely that Professor LePage helped Professor Healy make the decision to come to Rochester.

**Figure 7.** Daniel W. Healy.

Gerald Howard Cohen (Ph.D. Wisconsin, 1950) was appointed associate professor this first year to join Professor Healy; so also Assistant Professor Barry Dutcher (M.S. Rochester, 1958) and Lecturer William Coombs (M.S. Rochester). Professor Cohen had been employed by the Taylor Instrument Company in Rochester. He would organize and teach the first undergraduate courses in electronics and continue to carry these courses for many years thereafter as new individuals joining the department were given other teaching assignments. He also started graduate students on research problems in system dynamics and feedback control.

The first curriculum for students in the department was published in the 1958–59 Official Bulletin of the University. This shows thirty-two courses taken outside the department compared to nine given by the department.

**Electrical Engineering**

| | | | Hours | |
|---|---|---|---|---|
| | | | 1st Term | 2nd Term |
| *Freshman Year* | | | | |
| Phys. | 7 & 8 | Introductory Physics | 4 | 4 |
| ME | 1 & 2 | Engineering Drawing I and II | 3 | 3 |
| Math. | 9 | Analytic Geometry and Calculus | 3 | 3 |
| Engl. | 1 & 2 | English Composition | 3 | 3 |
| Electives | | Group II (Social Studies) | 3 | 3 |
| Ph. Ed. | 1 & 2 | Physical Education | 1 | 1 |
| | | | 17 | 17 |

| | | | | |
|---|---|---|---|---|
| *Sophomore Year* | | | | |
| Engl. | 3 & 4 | Introduction to Literature | 3 | 3 |
| Math. | 15 & 16 | Intermediate Calculus | 4 | 4 |
| Phys. | 11 & 12 | Physical Measurements | 4 | 4 |
| Chem. | 5 & 6 | General Chemistry | 4 | 4 |
| EE | 39 | Introduction to Circuit Analysis | 3 | |
| EE | 40 | Electrical Machinery I | | 3 |
| Ph. Ed. | 11 & 12 | Physical Education | 1 | 1 |
| | | | 19 | 19 |
| *Junior Year* | | | | |
| Electives | | Group II (Social Sciences) | 3 | 3 |
| Math. | 135 & 136 | Adv. Calculus and Applied Math. | 3 | 3 |
| Phys. | 121 & 122 | Electricity and Magnetism | 3 | 3 |
| Phys. | 124 | E and M Laboratory | | 1 |
| ME | 55 & 56 | Thermodynamics for EE | 3 | 3 |
| EE | 51 | Electrical Machinery II | 3 | |
| EE | 53 | Transients in Linear Systems | 4 | |
| EE | 64 | Engineering Electronics I | | 4 |
| | | | 19 | 17 |
| *Senior Year* | | | | |
| Electives | | Group I (Literature or Arts) | 3 | 3 |
| Electives | | Unrestricted | 3 | 3 |
| Phys. | 115 | Introduction to Atomic Physics | 3 | |
| Optics | 168 | Electronic Properties of Solids | | 3 |
| ME | 51 | Mechanics for EE | 3 | |
| Met. | 62 | Engineering Metallurgy for EE | | 3 |
| EE | 71 | Engineering Electronics II | 3 | |
| EE | 79 & 80 | Electrical Systems I and II | 3 | 3 |
| EE | 82 | Filters, Fields and Waves | | 3 |
| | | | 18 | 18 |

Students electing to major in electrical engineering were not admitted to the department until their junior year, thus continuing the tradition followed by the Department of Mechanical Engineering. Consequently, during their first two years in the College of Arts and Science, engineering students had time to pass preparatory courses and to be part of the liberal arts community of the university before specializing.

Additional members of the faculty were recruited quickly. In 1959, W. Richard Stroh (Ph.D. Harvard, 1957) and Hugh Guthrie Flynn (Ph.D. Harvard, 1956) arrived. Both had been at the Acoustic Research Laboratory at Harvard and were probably recruited through Professor Healy's personal contacts with the laboratory. He may also have seen the potential for recent Ph.D. recipients to develop acoustics research in a new department, although acoustics at that time was not often considered to be a part of electrical engineering. Hing-Cheong So (Ph.D. Illinois, 1960) was hired in 1960. In 1961, Edwin L. Carstensen (Ph.D. Pennsylvania, 1955), Daniel S. Ruchkin (Ph.D. Yale, 1960), and Herbert B. Voelcker, Jr. (Ph.D. London, 1961) joined the department. William Streifer (Ph.D. Brown, 1962) came in the fall of the following year. Professors So, Ruchkin, Voelcker, and Streifer saw opportunities to help shape a new department and brought with them recent research experiences in networks, communications, and fields. Professor Carstensen was invited for an interview as a result of a recommendation from Professor Herman Schwan, University of Pennsylvania, who with Professor Healy was a member of an early university grants committee of the National Institutes of Health.

In 1962, the faculty consisted of nine professors prepared to work in a variety of research areas:

- Professor Healy: microwaves, solid-state devices
- Associate Professor Carstensen: electrical and acoustical properties of biological material
- Associate Professor Cohen: servomechanisms, instrumentation and control
- Associate Professor Flynn: theoretical and experimental study of acoustic cavitation
- Associate Professor Stroh: architectural acoustics
- Assistant Professor Ruchkin: communications
- Assistant Professor So: networks
- Assistant Professor Streifer: electromagnetics, applied mathematics
- Assistant Professor Voelcker: communications, theory of modulation

Thomas Jerome Harris (M.S. Rochester), Instructor, and James Livingston Douglas (M.S. Rochester, 1960), Assistant Lecturer, were also listed with the department; both were part-time appointments.

This same year, the department was fortunate to benefit from an extraordinary award of one million dollars that the college received from the Alfred P. Sloan Foundation in New York City. Similar awards were given to Dartmouth College and to Brown, Johns Hopkins, and Princeton Universities.

The new faculty decided that the curriculum would reflect those areas of the discipline that were emerging at the expense of other areas that were being overshadowed. Older departments with committed faculty, courses, and laboratories could not abandon these latter areas easily, even though decreasing employment opportunities for new graduates in these areas were evident. For example, there would be very little if any classroom time or laboratory work given to electrical machinery or electrical power distribution.

Office and laboratory space for the new department was made available in Gavett Hall, the building that was already occupied by the Departments of Mechanical and Chemical Engineering. A wing had been added to Gavett Hall in the late 1940s when Professor Belknap was head of the Division of Engineering, and then a third floor was constructed on top of the wing in 1960. These additions made it possible to squeeze in electrical engineering and the department remained there until the Hopeman Engineering Building was completed in June 1963. The major portion of this new 30,000 square foot, $1.5 million building was contributed to the university through a bequest from the estate of Bertram C. Hopeman and a gift from Mr. and Mrs. Albert A. Hopeman, Sr.

Dr. Arthur W. Kantrowitz, physicist and director of the Avco-Everett Research Laboratory, was the principal speaker at the cornerstone laying for the new building. Objects placed in the cornerstone included a transistor, a circuit module of the type used in "modern computers," a cube of pyroceram such as is used in the nose cone of a rocket, and other materials pertaining to the state of the art of engineering circa 1962. Dr. John W. Graham Jr., dean of the College of Engineering, accepted the building on behalf of the university and later gave a presentation on "The New World of Engineering and Applied Science at Rochester" to alumni and other guests.

When the Departments of Electrical Engineering and Mechanical Engineering moved to the Hopeman Engineering Building, the vacated offices and laboratories provided expansion space for the Department of Chemical Engineering.

From 1958 to 1961, the undergraduate student enrollment in the College of Engineering increased from 250 to 300; the graduate enrollment went from 15 to 60. All three engineering departments, Electrical, Mechanical and Chemical, were qualified to award the Ph.D. degree. Over the same years, sponsored research increased six-fold, from $30,000 to $180,000. The Institute of Optics transferred from the College of Arts and Science in September 1961, and the college was renamed the College of Engineering and Applied Science. Professor Oscar E. Minor, Department of Mechanical Engineering, was the assistant dean. The average nine-month salary for professors was $10,530, and for assistant professors, $5,950; teaching and research assistant stipends were $1,700 to $2,000.

PROGRAM: HOPEMAN ENGINEERING BUILDING

| Administration | | | sq. ft. |
|---|---|---|---|
| 1. Reception and waiting area | Joint ME, EE | 300 | |
| 2. Clerical space (3 girls) | | 300 | |
| 3. Work room and supplies | | 100 | |
| 4. E. E. Department Chairman | | 200 | |
| 5. E. E. Asst. Dept. Chairman | | 150 | |
| 6. M. E. Department Chairman | | 200 | |
| 7. M. E. Asst. Depart. Chairman | | 150 | 1,400 |

Staff Offices

| | |
|---|---|
| Total of 26 private staff offices (18 E. E., 8 M. E.), varying in size from 120 to 200 sq. ft. (150 sq. ft. avg.) | 3,900 |

Seminar Rooms

| | |
|---|---|
| Four required @250 sq. ft. (consider locating these in pairs with ceiling-height movable divider between each pair.) | 1,000 |

Laboratories

| | | |
|---|---|---|
| 1. a. Basic electronic circuits lab. | 1,300 | |
| b. Microwave lab. | 1,300 | |
| c. Instrument room for above | 400 | |
| 2. a. Electricity and magnetism lab. | 1,300 | |
| b. Project lab. | 1,300 | |
| c. Instrument room for above | 400 | |
| 3. a. Elec. machine and heavy equipt. lab. | 1,600 | |
| b. Motor generator room | 400 | |
| c. Stock and instrument room | 500 | |
| 4. Communications systems lab. | 800 | |
| 5. Analog computer lab. | 800 | |
| 6. Staff research labs., 10 req'd. (6 E. E., 4 M. E.) @400 sq. ft. (*) | 4,000 | 14,100 |

Graduate Student Research-Study Areas

| | |
|---|---|
| 1. Total of 26 (14 E. E., 12 M. E.) research-study areas @200 sq. ft. (**). | 5,200 |
| 2. Total of 9 (5 E. E., 4 M. E.) study areas @300 sq. ft. (***). | 2,700 |

| | |
|---|---|
| Dark Room | 100 |

Lounge

| | |
|---|---|
| For staff and students; equipped with kitchenette | 1,000 |
| Receiving Room | 500 |
| Storage | 1,000 |
| Total net sq. ft. | 30,900 |

**Figure 8.** Preliminary plans for space usage in the Hopeman Engineering Building.

The university commencement program for 1958 has the names of four individuals receiving the Master of Science in Electrical Engineering degree; this graduation corresponded to the first academic year the new department existed. The names listed are Stanley J. Dudek, Lorand W. Magyar-Wilczek, Ronald G. Matteson, and Clor William Merle. At the next commencement, two more M.S. degrees in electrical engineering were awarded, to Joseph A. Huie and Peter Harold Zachmann. Miles Davis, James L. Douglas, and Clifford P. Oestreich also received this graduate degree in 1960.

The first class of undergraduates to receive the Bachelor of Science degree, major in Electrical Engineering, graduated in 1961:

| | |
|---|---|
| Elwyn G. Allyn | Donald G. Simcox |
| Lloyd R. Campbell, Jr. | Harold F. Staudenmayer |
| Francis J. Caravaglio | |

Nine Master of Science in Electrical Engineering degrees were also awarded in 1961:

| | |
|---|---|
| Arthur W. Alphenaar | David Platnick |
| George A. Brown | Thomas Proctor |
| Robert C. Curry | Raymond J. Rogers |
| Thomas E. Hattersley, Jr. | Menashe Simhi |
| Vladimir Kushel | |

Graduations from the department would continue in subsequent years. In 1962, seven men received the B.S. degree and eight the M.S. degree.

| | | |
|---|---|---|
| B.S. | Lee F. Backus | Gerald L. Freed |
| | Charles K. Bowman | Theodore H. Morse |
| | Marvin D. Clark | Kenneth G. Shepherd |
| | Thomas G. Coleman | |
| M.S. | Robert H. Aronstein | Theodore B. Mehlig |
| | W. Bromley Clarke | Arthur R. Phipps |
| | Michael K. Karsky | Frederick G. Reinagel |
| | Robert E. Lee | Robert A. Rubega |

At the 1963 graduation exercises:

| | | |
|---|---|---|
| B.S. | Michael F. Armstrong | Nicholas A. Milley |
| | James J. Ashton | Armando Scacchetti |
| | Alan Bernstein, Jr. | James E. Summers |
| | Gerald E. Claflin | James H. Taylor |
| | William J. McKechney | |
| M.S. | Bertrand E. Berson | Lee E. Ostrander |
| | Edward S.I. Chiang | Kodati S. Rao |
| | Anthony I. Eller | James R. Verwey |
| | J. Warren Gratian | Leon L. Wheeless, Jr. |
| | Robert F. Osborne | Walter W. White |

Parenthetically, Wilson Allen Wallis became the sixth president of the university in 1962 and continued in this office until 1970 when he was given the title of chancellor.

The University of Rochester, through the Department of Electrical Engineering, is known to be one of the first universities to have had a commitment to biomedical engineering; two others were the University of Pennsylvania and Johns Hopkins University. This early work in biomedical engineering within the department appears to be a direct result of two events that occurred in 1961. The first was Professor Healy's appointment of Lee B. Lusted (M.D. Harvard, 1950), Associate Professor of Radiology, as Professor of Biomedical Engineering, part-time, effective February 1961. Professor Lusted was instrumental in obtaining the department's first grant for biomedical engineering. This was from the National Institutes of Health for biomedical engineering graduate training. The second event was the arrival of Professor Carstensen the following September. Professor Carstensen had received one of the very early Ph.D. degrees in biomedical engineering while working with Professor Schwan at the University of Pennsylvania and

**Figure 9.** Edwin L. Carstensen.

was ready on arrival to start graduate instruction in biomedical engineering.

Although the Department of Electrical Engineering would carry the most active role in biomedical engineering in the college from this point on, the interdisciplinary field quickly spread to other departments.[3] As student interest developed in this new interdisciplinary area, it was decided within the college to continue to award existing engineering degrees only and not create a separate degree for biomedical engineering. The arguments against creating a special degree, arguments that would continue for many years, centered on unrealistic career expectations and the value of a traditional engineering degree to a young engineer entering a cross-disciplinary field. Instead, the departments in the college broadened their perspectives to include biomedical engineering problems as legitimate fields of research for graduate degrees.

By 1960, Dr. Lusted had published a number of papers on the potential use of digital computers in the medical field, including "Computer Programming of Diagnostic Tests"[4] and "The Use of Electronic Computers in Medical Data Processing: Aids in Diagnosis, Current Information Retrieval, and Medical Record Keeping."[5] Professor Cohen along with D. Platnick, had also published a paper, "Design of Transistor IF Amplifiers Using an IBM 650 Digital Computer."[6]

Seminar courses in biomedical engineering were offered during the 1961–64 academic years to focus attention on active areas of research. Invited speakers were major figures in biomedical engineering in the country at that time. Six research areas and their representatives included: 1) dielectric properties of biological material: Herman Schwan and Gerhardt Schwartz, Pennsylvania; and Professor Carstensen; 2) computational methods in neurophysiology: Stanford Goldman, Syracuse; Moise Goldstein, Wesley Clark and Jerry Lettrin, MIT; Heinz von Foerster, Illinois; E. Roy John, Rochester; 3) cardiovascular systems: Samuel Talbot, Johns Hopkins; Richard McFee, Syracuse; David Robinson, Johns Hopkins; Fred Grodins, Northwestern; L. Peterson, Pennsylvania; H. Warner, Utah; Paul Yu, Arthur Moss and Carl Honig, Rochester; and Professor Kinnen; 4) biological control systems: Manfred Clynes, Rockland State Hospital; Stanley Molnar, Epsco Medical, Cambridge, Mass.; P. Willis and Lawrence Stark, MIT; J.F. Gray and R. Jones, Northwestern; I. Lowenfeld, Columbia; D. Robinson, Johns Hopkins; J. Mundie, Wright-Patterson Air Force Base; Lawrence Goodman, Case Institute of Technology; Jan Stolwijk, John B. Pierce Foundation, New Haven, Connecticut; Robert Boynton and R. Abrams, Rochester; and Professor Cohen; 5) medical instrumentation: Stuart MacKay, California; Edwin Gordy, Roswell Park, Buffalo, N.Y.; Douglas Parkhill, General Dynamics, Rochester; Scott Swisher and David Wilson, Rochester; 6) audition: Edward David, Bell Telephone Laboratories; James Zwieslocki, Syracuse; R. Galombus, Yale; P. Coleman, Maryland; Karl Lowy, Rochester; and Professors Ruchkin and Voelcker; 7) biomedical ultrasound: Wesley Nyborg, Vermont; Eugene Ackerman, Mayo Clinic; Jack Reid, Pennsylvania; John Jacobs, Northwestern; William Fry, Illinois; Justus Lehman, Washington; Henning von Gierke, Wright-Patterson Air Force Base; and Professors Carstensen and Flynn.

There is no recollection that Professor Healy had an express intent to make acoustics a special research interest when he became chair of the department. But as mentioned, acoustics became a part of the department when Professor Stroh arrived with his interest in acoustic research. Professor Stroh then helped to attract Professor Flynn who was working in physical acoustics, specifically in cavitation phenomena. When Professor Stroh left in 1962 to go to Goucher College, Maryland, Professor Flynn encouraged Professor Healy to interview David T. Blackstock (Ph.D. Harvard, 1960) from the same Harvard Acoustics Laboratory, who happened to be working in Rochester at the General Dynamics/Electronics Corporation. Professor Blackstock joined the faculty in 1963. Because of the difficulty of attracting graduate students to acoustic research at the University of Rochester, Professor Blackstock resigned in 1970 to take a position at the Applied Research Laboratories, University of Texas at Austin.

**Figure 10.** David T. Blackstock.

However, acoustic research in the 1960s in the department was notably productive. In 1964, Professor Flynn authored a seminal review of acoustic cavitation that was to become a major reference for subsequent cavitation research in this country and abroad. Years later, he would collaborate with Professor Carstensen when the latter encountered

**Figure 11.** Faculty, graduate students and families at Webster Park, Rochester, New York, late Spring, 1964.

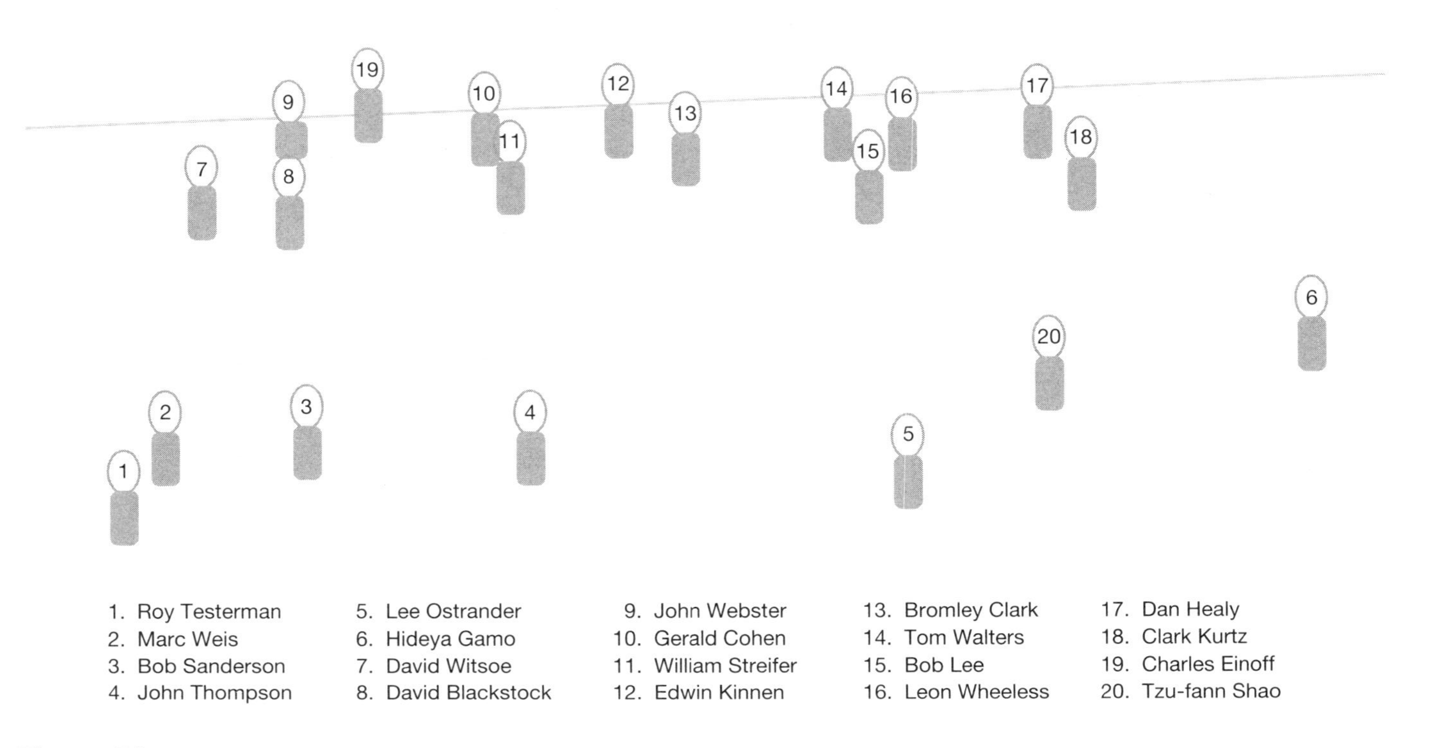
1
2
3
4
5
6
7
8
9
10
11
12
13
14
15
16
17
18
19
20
1. Roy Testerman
2. Marc Weis
3. Bob Sanderson
4. John Thompson
5. Lee Ostrander
6. Hideya Gamo
7. David Witsoe
8. David Blackstock
9. John Webster
10. Gerald Cohen
11. William Streifer
12. Edwin Kinnen
13. Bromley Clark
14. Tom Walters
15. Bob Lee
16. Leon Wheeless
17. Dan Healy
18. Clark Kurtz
19. Charles Einoff
20. Tzu-fann Shao

**Figure 12.**

cavitation phenomena in living tissue. During his period in the department, Professor Blackstock wrote several papers on nonlinear acoustics, two of which were recognized years later with major awards from the Acoustical Society of America.

Three more individuals received appointments to the department during the summer of 1963. Lloyd P. Hunter (D.Sc. Carnegie Mellon, 1942) came from the IBM Research Center in Yorktown Heights, N.Y., where he had been the director of component engineering and manager of solid-state component development. Hideya Gamo (Ph.D. Tokyo, 1958) also came from the IBM Research Center in Yorktown Heights as a visiting professor. His interests were in the coherence properties of light and in information theory applied to optical problems. Edwin Kinnen (Ph.D. Purdue, 1958) transferred from the University of Minnesota, attracted by opportunities to contribute to the biomedical engineering program while continuing his research interests in system dynamics. These new people compensated on the teaching side for the departure of Dr. Lusted and Professor Stroh. The department now had five major research areas: Acoustics, Biomedical Engineering, Communications and Information Processing, Solid State Electronics, and Systems and Control.

During the next six years, the faculty increased gradually from ten to thirteen. Professor Gamo's appointment was changed from visiting professor to professor. Edward L. Titlebaum (Ph.D. Cornell, 1964) in effect replaced Professor Ruchkin in 1964 when Professor Ruchkin went to the Brain Research Laboratory at the New York Medical College. Michael M. Reid (Ph.D. California) had an appointment as assistant professor, part-time, in the department for a few years while he was attending medical school at the university. Professor So left for the Bell Laboratories in September 1965.

Then in 1967 and 1968, three more appointments were made: Pankaj K. Das (Ph.D. Calcutta, 1964), Robert C. Waag (Ph.D. Cornell, 1965), and Sidney Shapiro (Ph.D. Harvard, 1959). Professor Das had been with the semiconductor group at the Polytechnic Institute of Brooklyn and Professor Waag with the Communication Division of the Rome Air Development Center of the US Air Force. Professor Shapiro came from the Bell Laboratories. Earlier, at Arthur D. Little, Inc., he had provided the first experimental confirmation of the ac Josephson effect in superconducting thin film junctions. He discovered, when these junctions are irradiated with microwaves, that sharp rises in current appeared at fixed voltages in the current-voltage characteristic of the junction. This effect, often labeled "Shapiro steps" in the literature, was later developed by others, including his graduate student, Clark Hamilton, to serve as a voltage standard. The existing international voltage standard is based on this phenomenon.

**Figure 13.** Gerald H. Cohen and student with apparatus for measuring pupillary response, in a visual science laboratory, 1967.

During the 1960s, Professor Cohen had started a research program in cooperation with Robert M. Boyton, Director of the Center for Visual Science, and faculty members of the Department of Ophthalmology. Eye movement and visual perception were of interest initially. Leon Wheeless, who had been working with Professor Cohen, submitted his Ph.D. dissertation in 1965 on "The Effects of Intensity on the Eye Movement Control System." At about the same time, Julius Goldstein, studying with Professor Voelcker, submitted his thesis titled "An Investigation of Monaural Phase Perception." These two dissertations would be the first in a long line that would emerge from biomedical engineering studies within the department. These theses were also studies in biomedical engineering that evolved directly from conventional areas of electrical engineering, the former from feedback control systems and the latter from communication theory. Adding to the department's activity and visibility in biomedical engineering studies, Professor Kinnen began working with Paul Yu, Department of Cardiology, investigating the possibility of measuring thoracic blood flow and cardiac output based on changes in a high-frequency electric field placed across the human thorax. This in turn led back to a more engineering-oriented problem, that of calculating an electric field in a complex three-dimensional geometry with many internal boundary conditions. This calculation was done by inverting a 180,000 node matrix on a CDC 6600 computer located at New York University, as there was insufficient computing power on the Rochester campus.

The focus of Professor Carstensen's research during these early years was on the high frequency acoustic properties of tissues and cells and on bioelectricity. The latter and the discovery of the pervasive dominance of bound charge on the dielectric properties of microorganisms led to a collaboration with Robert Marquis, Department of Microbiology, that continued into the 1980s. Professor Carstensen would have National Institutes of Health support for his research on the dielectric properties of biological materials and bioultrasound over the next thirty-five years, an unusual length of time for any one investigator to receive support from the NIH. Sally Zehr Child joined the department as a laboratory technician in 1965 and assisted Professor Carstensen with his research until his retirement.

The beginning of bioacoustic or bioultrasound research at the university was not anticipated or planned. Rather it developed simply from individual researchers in disparate fields making contact within the confines of a small university. Nevertheless, by the end of the 1990s, the university would include what was probably the world's largest group of professionals devoted to the applications of ultrasound to medicine and biology. As noted earlier, the department had had a program in physical acoustics from its earliest days. The work in bioultrasound, however, began with a meeting between Professor Carstensen and Raymond Gramiak, Department of Radiology. During the 1960s, Dr. Gramiak had been using ultrasound clinically in the evaluation of heart function. He observed, when an x-ray contrast agent was injected into the heart, that a very large echo was produced in the ultrasound image. During a visit, Douglas Gordon, a radiologist from England and the author of the first published book on medical ultrasound, suggested to Dr. Gramiak that this echo might be due to bubbles caused by the injection. He also suggested that Dr. Gramiak contact Professor Carstensen, who Dr. Gordon knew had published some work in this area. Professor Carstensen had studied the scattering and attenuation of ultrasound by bubbles during World War II prior to coming to Rochester. Together Dr. Gramiak and Professor Carstensen formulated three alternative hypotheses to explain the echo, among them the bubble hypothesis. Fred Kremkau, then a graduate student interested in acoustics, became part of this early collaboration and confirmed the bubble hypothesis.

Research in bioultrasound has been a characteristic of the department ever since. Completing his Ph.D. dissertation, "Macromolecular Interaction in the Absorption of Ultrasound in Biological Material," Dr. Kremkau would continue his interest in bioultrasound, eventually becoming the director of the Medical Ultrasound Center at the Bowman Gray School of Medicine at Wake Forest University and president of the American Institute of Ultrasound in Medicine. Other early Ph.D. degree recipients who had carried out their thesis research in biomedical engineering were Marc Weiss, who became chief scientist at the Naval Biodynamic Laboratories in Louisiana, and W. Bromley Clarke, who went on to become vice president and head of Research and Development of Johnson & Johnson Medical Supplies.

Professor Waag began a long career at the University of Rochester also in bioultrasound. As a new member of the faculty looking for a viable research area, Professor Carstensen suggested that he talk to Dr. Gramiak about the possibility of extending his background in communications and radar to the development of diagnostic ultrasound. This led to a congenial and dynamic collaboration that would continue until Dr. Gramiak's retirement in 1988. Their research on ultrasonic scattering, propagation, and medical imaging took place in a nearby laboratory in the Strong Memorial Hospital where they would be close to patients. Consequently, they were very early to use computers to assist diagnoses in a medical clinic. Publications of research results were about contrast agents and the medical use of ultrasound for differentiating malignant tissue from healthy tissue by the analysis of the differences in reflected ultrasound waves. This collaborative work was funded in part by the National Science Foundation and the National Institutes of Health for over thirty years. Along the way, Professor Waag established an Industrial Associates Program for the Diagnostic Ultrasound Research Laboratory.

**Figure 14.** Robert C. Waag.

Corporate members of this program included Eastman Kodak Company, Fujitsu Laboratories, General Electric, Hewlett-Packard, Hitachi Medical, Phillips Laboratories, Siemens, and Hoffman LaRoche. Professor Waag had a joint appointment with the Department of Radiology since 1973.

Device physics and superconductivity were other areas of research during the decade of the 1960s. Investigations would result in the

**Figure 15.** Lloyd P. Hunter and students.

development of new and novel semiconductor devices, high temperature superconductive material applications, and ultra high-speed phenomena. Publications describing these studies often appeared before similar studies were started in other departments of electrical engineering around the country. When the Laboratory for Laser Energetics (L.L.E.) was established in 1970, some of the work in device physics was transferred to that laboratory by faculty of the department who chose also to be part of the facility. A timely event occurred in 1969 when the department and Consolidated Vacuum Corporation in Rochester jointly sponsored a symposium on the deposition of thin films by sputtering techniques. Fourteen industrial laboratories were represented at the symposium, four universities, three governmental agencies, and scientists from five countries.[7]

Professors Gamo and Streifer had early support from the National Science Foundation and the U.S. Air Force Office of Scientific Research to study optical wave-guides and the statistical properties of laser radiation. Part of the joint research was to build a high-gain infrared gas laser, Xenon 3.5 microns. Toshiharu Taku (D.Sc. Tokyo) joined the project as a visiting research associate to assist in building the ultra-high vacuum system required by the laser. He was on a sabbatical leave from the Institute of Metrology, the Ministry of International Trade and Industry of Japan. When Dr. Taku returned to Japan in 1967, Shigeo Asami (Ph.D. Tokyo) would spend a postdoctoral year in the department working with the laser that had become functional. Shortly thereafter, however, Professor Gamo transferred his work and laboratory to the University of California at Irvine.

The late 1960s were known for their social unrest and the Vietnam War. These were the years of anti-war protests and the assassinations of Martin Luther King, Jr., and Robert F. Kennedy. College campuses, including Rochester, were set into uproar after the shooting of students at Kent State University in Ohio by National Guard troops called out to control demonstrations. Nevertheless, the department's undergraduate enrollments held steady at about 100. In contrast, the graduate enrollments fluctuated between 50 and 90. An average of 13 students graduated with a B.S. degree, with a one-year maximum of 20. Masters degrees averaged 17, Ph.D. degrees about half of this. Of the total of 26 Ph.D. degrees granted to students of the department by the end of 1969, 10 of the dissertations were in biomedical engineering, 8 in physical and high frequency phenomena, and 8 in communications and systems. Publications by members of the faculty were averaging 16 per year.

In an effort to reach out further into the community, the college began a program in the mid 1960s leading to an M.S. degree in General Studies with a designated field of concentration. This program was intended for part-time students, those individuals working full-time locally who were interested, for example, in recent developments in communication theory, quantum electronics, or biomedical engineering. At this time, the department received permission from the Graduate Council of the University to invite an individual from outside the university to participate as a member of the examining committee with full vote on the occasion of a Ph.D. thesis defense. This was done both as self-discipline and to inform key people in local industry and other universities of the quality of the research in this new department.

Interesting in retrospect, the department had a policy in the 1960s of discouraging its undergraduates from remaining at Rochester for graduate study. It was felt that the student would benefit from continuing his or her studies with a different faculty in a different academic environment. However, during the early years of the Vietnam war, Congress passed the Military Selective Service Act of 1967 that changed the draft law. Students would no longer be deferred past age twenty-four or after four years of college. Graduate student applications dropped sharply due to this change and the policy disappeared simply as a practical matter.

In the meantime, Dean Graham left the University to become president of Clarkson College of Technology, Potsdam, N.Y. Professor Robert G. Loewy (Ph.D. Pennsylvania, 1962) from the recently renamed Department of Mechanical and Aerospace Science was appointed the new dean, although Cecil E. Combs substituted as acting dean during the 1966–67 academic year. Professor Loewy came to the University of Rochester in 1962 from the Vertol Division of the Boeing Company where he was a recognized authority on helicopters. Dean Combs had been a Major General in the U.S. Air Force, retiring in 1965. The associate dean for Graduate Studies was Professor Flynn and the assistant dean was Carlyle F. Whiting (M.S. RPI, 1950). Dean Whiting was a graduate of the U.S. Military Academy, 1944, and a fighter pilot in the Army Air Corp, retiring as a Lieutenant Colonel.

**Figure 16.** Hugh G. Flynn.

Following the untimely and tragic death of Professor Dan Healy in 1969, Professor Flynn was appointed acting chair of the department. Professor Healy's time as chair was in the postwar years when physicians and scientists were just beginning to recognize the potential contributions technology could make to many branches of academic medicine. Professor Healy took a chance during his ten-year tenure to allow the department to move into biomedical engineering. As a result, biomedical engineering research became a critical aspect of the department's research, with a majority of the members of the faculty becoming involved over the next thirty years.

Sabbaticals and leaves were granted during these first years. Professor Voelcker went to Imperial College, London, England, on a NATO postdoctoral fellowship during the 1967–68 academic year, the same time that Professor Hunter was on leave at the Research Laboratory of the IBM Corporation in San Jose, CA. Professor Voelcker had received the Edward Peck Curtis, Jr., Award for Excellence in Undergraduate Teaching in 1969. Professor

Flynn took a sabbatical leave in 1968–69 at the Acoustic Research Laboratory at Harvard University and the University of Cambridge, England. The next year, Professor Cohen was at the University of Southern California, Professor Streifer at Stanford University, and Professor Blackstock at the University of Texas.

A number of colleagues from other departments of the university were helpful during these formative years of the department, assisting both faculty and graduate students. Many of these individuals were participants in the early biomedical engineering seminar series. Others in particular included: Stanley F. Patten, Department of Pathology; Henry S. Metz, Department of Ophthalmology; Raymond S. Snider, Department of Anatomy and the Center for Brain Research; Marylou Ingram, Department of Radiation Biology and Biophysics; Jack Maniloff, Department of Microbiology; Robert E. Hopkins, Institute of Optics; Joseph H. Eberly, Emil Wolf, and Leonard Mandel, Department of Physics and Astronomy. Here again, the small size of the university and the proximity of the medical school facilitated these interactions.

Faculty research attracted some overseas visitors during this time. Amos Nathan, Professor of Electrical Engineering at the Technion in Haifa, Israel, was a visiting professor during 1969–70. Visiting Scientists included Vytaly Dmitriev and Constantin Khromenko from the Low Temperature Institute, Kharkov, Ukraine, U.S.S.R. John B. Waugh (M.S. New South Wales) stayed as a part-time Senior Research Associate during the years 1964–1968.

At the close of the 1969–70 academic year, the faculty numbered twelve plus one visiting appointment.

- Acting Chair Professor Hugh Flynn
- Professors Gerald Cohen, Lloyd Hunter
- Associate Professors David Blackstock, Edwin Carstensen, Pankaj Das, Edwin Kinnen, Sidney Shapiro, William Streifer, Edward Titlebaum and Herbert Voelcker
- Assistant Professor Robert Waag
- Visiting Professor Amos Nathan

W. Allen Wallis was President of the University; Robert Loewy, Dean of the College of Engineering and Applied Science; Hugh Flynn, Associate Dean for Graduate Studies and Acting Chair of Electrical Engineering; Carlyle F. Whiting, Assistant Dean.

**Figure 17.** Paul Osborne, right, Senior Technical Associate, and student, with an ion implanter donated to the department by the Eastman Kodak Company, ca. 1985.

# 4: The 1970s

"Nothing exists from whose nature some effect does not follow."
*Spinoza*

The department was known twelve years after its founding for significant research efforts in biomedical engineering and biomedical ultrasound, communications and information processing, solid-state electronic devices, and superconductivity. Collaborations in biomedical engineering were occurring with four departments in the School of Medicine and Dentistry; and there was grant support from the National Science Foundation, the National Institutes of Health, and the U.S. Air Force Office of Scientific Research. The National Science Foundation and the U.S. Navy were funding communications research. Department research expenditures totaled $242,198 for the 1971–72 fiscal year. In addition to those sources mentioned research money was also received from the Taylor Instrument Company, General Telephone Electronics, and New York State Science and Technology.

The early 1970s were years when significant changes were occurring within the engineering academic communities. This chapter describes some of these changes and discusses the responses of the college and the department to these changes.

Digital computers were becoming widespread and new communication systems were being adopted. Devices and products using solid-state electronics were appearing everywhere. All this, it seems, should have led to renewed interest on the part of high-school graduates for careers in electrical engineering. Unfortunately such an interest did not occur as this was also the time when young people were questioning the social values of technology and skeptical of the job opportunities for engineering graduates. The result was fewer undergraduate enrollments in engineering throughout the country. The number of federal and state fellowships was also declining and this placed pressure on members of the faculty to seek increased graduate student support from government agencies as well as directly from industry.

Most engineering schools would attempt to attract the decreasing numbers of engineering undergraduates with variations to traditional programs. At the University of Rochester, requirements for a degree in electrical engineering were modified to allow students to choose four out of five basic electrical engineering sequences and then select an area of concentration. The two-course sequences were in communications, computer engineering, fields and waves, solid-state electronics, and systems engineering. The concentration required advanced courses to be taken in one of these areas. Biomedical engineering and environmental acoustics were also offered as advanced courses. Students were encouraged to take graduate-level courses as instructors made an effort to remove real and perceived boundaries between undergraduate and graduate courses. Opening graduate courses to undergraduates provided greater flexibility both in planning student programs and in scheduling elective courses. The latter could be offered every other year and still be available to all undergraduates. These innovations and others allowed faculty members to continue to schedule blocks of time for research projects. The innovations also proved sufficient at the University of Rochester for enrollments in the department's undergraduate courses to continue strong and in some cases to increase.

An Interdepartmental Degree Program was introduced in the college in 1971, also as a response to changed student interests and expectations. It was thought that an interdepartmental program would attract incoming students who had at least some interest in engineering. Each undergraduate in the program could create his or her own curriculum within broad constraints and carry out a required senior project. Many individuals with interests in medicine, law, business, or in biomedical engineering chose this route with faculty advisors in electrical engineering. As many as twenty-four interdepartmental students would work with department faculty during the early years of the program. This was a sizable number when compared to the 125 undergraduates at the time registered in electrical engineering. The college also had a 3-2 program whereby students could receive two degrees in five years. A 3-2 student could obtain a B.S. degree in electrical engineering plus a second degree, for example, from the College of Arts and Science or the School of

Business Administration or the Eastman School of Music, or possibly an M.S. degree in electrical engineering.

A new course, Insights Into Engineering, was offered college-wide about the same time for all first-year engineering students. Registrants were required to take three five-week short courses that introduced some aspect of engineering. Three were offered initially by department faculty: Noise Pollution in the Urban Environment with Professor Flynn, An Introduction to Echolocation Systems with Professor Titlebaum, and An Introduction to Fortran Programming with Engineering Applications with Professor Waag. Later, Professor Merriam would offer one on numeric computing using a powerful software product called APL.

The early 1970s was also the time when student interest in electronic musical instruments and sound reproduction was rampant. Professor Kinnen responded by offering an experimental course on things electrical for students with no engineering background. Officially called Electronics for Everyman, most of the fifteen students who elected the first offering of this course were juniors and seniors in the College of Arts and Science. None was a hard-science major, although at the end of the semester a number said that they regretted not having elected engineering on entering the university. The course was given only twice, as there were other instructional needs with higher priorities.

Coincident with changing student interests and new teaching initiatives, faculty members throughout the university were beginning to recognize the potential impact of digital computers on most if not all academic disciplines. This culminated in 1974 with the creation of a new Department of Computer Science on campus. Digital computers had been visible around the University since 1955. A Burroughs E101 had been installed in March 1956, then an IBM 650, and subsequently in 1961 an IBM 7070 and a 1401. Thomas A. Keenan (Ph.D. Purdue, 1955) was director of the University Computing Center that was functioning as a computing facility and service for the entire university community. Center personnel offered short non-credit courses in programming and associated topics as early as 1956. In 1958, Dr. Keenan proposed a graduate program in applied mathematics that would include courses in numerical analysis and computer applications. This program was offered within the university for a few years but did not develop further due to general lack of faculty support. Dr. Keenan resigned in 1966 and Vincent Swoyer became the new director of the Computing Center.

Nevertheless, computer courses, programming courses, and allied subjects did develop ad hoc over the late 1960s in the department, the Institute of Optics, the School of Business Administration, and in the Department of Mathematics. Three graduate-level courses in the department were EE 401 Computer Electronics, EE 492 Numerical Algorithms, and EE 493 Digital Signal Processing.

In 1969 Robert L. Sproull arrived from Cornell University as the new university provost. He had seen a Computer Science Department established at Cornell a few years earlier and recognized the possibilities for creating one at the University of Rochester. Dean H. Meckling, School of Business Administration, and Dean Loewy, particularly with the encouragement of Professor Flynn, then acting chair of Electrical Engineering, and Professor Voelcker, recommended that Provost Sproull establish a formal computer science department. A Committee on Computer Science was formed in January 1970, with Professor Flynn as chairman and Professor Voelcker and six other faculty from various departments as members. Working with outside consultants, the committee made a strong case for establishing a Department of Computer Science and recommended that a separate department be established immediately. The recommendation also stated that the new department should be outside the normal college structure to avoid whatever constraints, real or imagined, would exist if this new department were to be assigned either to the College of Arts and Science or the College of Engineering and Applied Science. The board of trustees approved the committee's report and a second committee was appointed to recruit a chairperson of the new department. Professor Voelcker was asked to chair this committee. It duly selected Jerome A. Feldman (Ph.D. Carnegie Mellon, 1964), Stanford University, for the position, and he came to Rochester in 1974. He brought new interests in the developing fields of programming languages, artificial intelligence, and distributed computing. Until he left the university in 1989, Professor Feldman held a joint appointment with the Department of Electrical Engineering.

The first undergraduate computer course in a curriculum at the university was given by Professor Voelcker in 1970. This course, EE100 Introduction to Computers, was intended for first and second year engineering students. The syllabus included binary notation and Boolean algebra, and an introduction to logic circuits, number systems, machine language, assembler code and elementary compiler theory. Computer experience was provided through simulations using a virtual machine executed on a mainframe machine in the Computer Center. Both men and women undergraduates throughout the university quickly discovered the course; ninety-three registered the first year of which only about half were from the College of Engineering and Applied Science. The second year Professor Voelcker taught the course, the enrollment well exceeded one hundred.

Charles W. Merriam III (Sc.D. MIT, 1958) came to the university in 1971 from Cornell University to become department chair, a position he would hold for the next nine years. Although his interests had been in control systems and optimization techniques, Professor Merriam would be instrumental in establishing a number of undergraduate computer engineering courses in the department and bringing recent Ph.D. degree graduates to the faculty to assist in developing the field. Programming courses and a logic design laboratory were quickly organized with the assistance of Professor H.C. Torng from Cornell, who spent his sabbatical year, 1973–74, working with Professor Merriam. Early during his tenure as chair, a new undergraduate laboratory was funded for computer architecture and another for solid-state electronics.

In the meantime, Professor Moshe Lubin (Ph.D. Cornell, 1966), Department of Mechanical and Aerospace Science, had begun experimenting with powerful lasers in a laboratory in the basement of Gavett Hall. By 1970, these experiments would develop into the Laboratory for Laser Energetics with Professor Lubin as its first director. Professor Lubin's laboratory in the basement of Gavett Hall was soon moved to the basement of the Hopeman Engineering Building. Faculty offices of both the Departments of Electrical Engineering and Mechanical and Aerospace Science were located around the perimeter of the three upper stories of this building. The central areas contained student laboratories and research areas. When concern was expressed to Professor Lubin about just what was being imploded or

**Figure 18.** Charles W. Merriam.

exploded down below, he would reply that he was just experimenting with energy levels equivalent to that found in a chocolate candy bar. However, anyone operating sensitive electronic equipment or oscilloscopes in other parts of the building soon learned to accept large pulsed interference whenever a laser in Professor Lubin's laboratory was triggered. These pulses were readily transmitted from the basement laboratory throughout the building over the electrical power wiring.

The function of the L.L.E. was to study laser fusion, high-density physics, and ultrafast science with the ultimate and formidable goal of demonstrating the feasibility of using high-energy pulsed lasers to achieve controlled thermonuclear fusion, that is, inertial confined fusion energy. As the L.L.E. expanded over a ten-year period, it gained worldwide recognition, in part, as a National Laser User Facility. Professor William Streifer was an early participant in the L.L.E.; however he would leave in 1972 to join the Xerox Palo Alto Research Park on the west coast.

As to department activities in biomedical engineering, Professor Carstensen began collaborative research with Morton Miller, Department of Radiation Biology, on the bioeffects of electric, magnetic, and ultrasonic fields. Professor Miller had redirected his research in the early 1970s from the effects of ionizing radiation on plant roots to the effects of electric fields at the 60-hertz power line frequency. Professor Carstensen would later complete a monograph summarizing the influence that his work and that of other investigators had had on controversies over the safety of fields generated by the transmission lines of the electric utility industry (see chapter 5).

In the mid 1970s, Charles Linke, Department of Urology, contacted Professor Carstensen to explore the possibilities for thermal surgery in the kidney using microwaves. They decided instead that ultrasound had a number of advantages over microwaves for the purpose, and this added another research dimension to biomedical engineering at Rochester. A series of studies demonstrated that tumors could be destroyed bloodlessly and without enhancing metastasis. However, while lesions could easily be produced when the kidney was in direct contact with the ultrasound source, the lesions could not be produced at the highest powers available when the kidney was at the focus of the transducer. Without knowing it, the investigators had encountered the phenomenon of nonlinear acoustic saturation familiar to scientists in the fields of underwater sound and airborne acoustics. Interesting in retrospect, Professor Blackstock had initiated his study of nonlinear acoustics while at the University of Rochester in the early 1960s and then continued the work at the University of Texas at Austin. After exploring the kidney observations with Professor Blackstock, Professor Carstensen began collaborating with Tom Muir, Blackstock's colleague at the University of Texas at Austin. This collaboration concluded with the introduction of the idea of nonlinear ultrasound propagation in tissue. Professor Blackstock became a member of the biomedical ultrasound program at the University of Rochester after this and was a summer resident in Rochester for a number of years as a visiting professor. David Bacon from the National Physics Laboratory in England also contributed significantly to the department's research program on nonlinear propagation in the 1980s and 1990s.

Additionally, Professor Titlebaum was studying ultrasonic waves, in this case ultrasonic waves used by several species of bats and by dolphins and whales for navigation. Earlier work on very broadband signals and ambiguity functions led to an investigation of Doppler tolerant effects using the high frequency sonic signals emitted by these animals. With these sonic signals, animals are capable of detecting objects independent of their own velocity. Thus bats, for example, are able to see the wall of a cave or an insect during complex flight maneuvers. By combining what some might consider intimidating engineering concepts with observations made of livings things, Professor Titlebaum demonstrated again the value of life-science studies as an element of engineering research. The flow of information in Professor Titlebaum's investigation, however, was more from the biology side of an engineering application. A notable paper on Doppler tolerant waveforms was written by Professor Titlebaum and his student Richard Altes.[1] Professor Titlebaum had U.S. Navy research support for over thirty years, some through the Naval Undersea Center at San Diego, the Applied Research Laboratory at Pennsylvania State University, and the Naval Undersea Warfare Center at New London, Conn., and Newport, R.I. He also received support later on from the Ballistic Missile Defense Organization (Star Wars) for radar studies and the use of frequency hop codes in military communications.

Of note, the Radiological Society of North America Scientific Assembly and Meeting in 1972 and again in 1975 presented the *Magna Cum Laude* Scientific Exhibit Award to Professor Waag and Dr. Gramiak, Department of Radiology. The first was for Cine Ultrasound Cardiography and the second for Tissue Macrostructure from Ultrasound Scattering. These distinctive awards helped to call the attention of the

**Figure 19.** Edward L. Titlebaum.

**Figure 20.** Herbert B. Voelcker, Jr.

international community to the theoretical and clinical studies of bioultrasound at the University of Rochester.

Other research activities were similarly increasing the visibility of the department. Professor Voelcker launched the Production Automation Project (P.A.P.) in 1972; it would continue at Rochester for the next fifteen years. This was an innovative program to devise new computer theories and applications for economically programming numerically controlled machines. Initially the work concentrated on expanding the field of solid modeling. Major funding came from the National Science Foundation program for Research Applied to National Needs. Aristides Requicha (Ph.D. Rochester, 1970) joined the project within a year. Early on, the P.A.P. had a collaborative relation with the Gleason Corporation as well as support from Xerox and General Motors. This support expanded into an Industrial Associates Program that eventually included McDonald Douglas, Boeing, Calma, IBM, Ford, Sandia National Laboratories, Digital Equipment Corporation, Eastman Kodak Company, Deere, Lawrence Livermore, Tektronix, and Westinghouse, in addition to Gleason, Xerox, and General Motors. Eugene Hartquist, Richard Marisa, and Professor Gershon Kedem, Department of Computer Science, were to join this group in subsequent years.

Superconductivity studies in the department emphasized the high-frequency properties of the Josephson effect in superconductor-insulator-superconductor (SIS) tunneling sandwiches. Proof was obtained to confirm the predicted peak in the superconducting energy gap. Other experiments demonstrated that Josephson junctions could function as high frequency devices for signal detection, mixing, and amplification. The superconductivity group became heavily invested in international collaboration. There were visits to a number of foreign laboratories, primarily in Scandinavia and the former Soviet Union, and several visiting scientists were accommodated in the Rochester laboratories for short and long stays. Among those who visited and contributed to the investigations were Mogens Levinsen, University of Copenhagen, Ole Hoffman Soerensen, Technical University of Denmark, Alexander Vystavkin, Institute of Radio Engineering and Electronics, Moscow, and Konstantin K. Likharev, Moscow State University. These and others were encouraged and given aid to make repeat visits, often to participate in scientific conferences. But it was still the era of the Cold War and on occasion travel to the West was denied to certain scientists from the U.S.S.R. Professor Shapiro volunteered to summarize their work at such conferences, aided by slides and manuscripts of those denied travel but brought by the one or two scientists allowed to travel.

Interestingly, while visiting Rochester and elsewhere in the U.S., Professor Likharev produced one of the first books on the Josephson effects and then translated it into English for publication in the U.S. Later, dissatisfied with policies in the Soviet Union, he emigrated to the U.S. As a professor at the State University of New York at Stony Brook, he continued to make substantial contributions to the development of circuits that use Josephson junctions for high-speed computing and data gathering and maintained collaboration with the department.

Research proceeded as well in other areas—those areas found more frequently in older Departments of Electrical Engineering. Professor Hunter had students working in semiconductor silicon. One of these students was Joan Ewing, the first woman to complete an advanced degree in electrical engineering at the University of Rochester. She received an M.S. degree in 1967 and a Ph.D. degree in 1973. Statistics compiled by the Society of Women Engineers show that only 0.8 percent of bachelor's degrees awarded in 1971 in all of the engineering disciplines went to women. Also, only 1.7 percent of earned engineering doctorates in 1975 were granted to women.[2]

Ms. Ewing had a successful career in Rochester with the Xerox Corporation, conducting studies on the electrical properties of materials at the Joseph C. Wilson Center for Research Technology in Webster. When she started with the company in 1956, it was known as

**Figure 21.** Joan Ewing.

The Haloid Company, the small firm that grew eventually into a Fortune 500 corporation.

During the 1970s Professors Cohen and Kinnen were publishing technical papers on system optimization and nonlinear system stability, and Professor Waag was investigating some aspects of digital communication systems. Professor Cohen, with a joint appointment in the Departments of Electrical Engineering and Ophthalmology, and also a member of the Center for Visual Science, continued his vision research concentrating now on the visual tracking system.

On the administrative side, the years 1974–75 found W. Allen Wallis elevated to the new position of chancellor of the university. Robert L. Sproull became the president and a new dean of the College of Engineering and Applied Science had to be appointed. Dean Loewy left early in 1974 to become vice president and provost at the Rensselaer Polytechnic Institute. Rudolf Kingslake (D.Sc. London, 1950), Institute of Optics, was named acting dean until Brian J. Thompson (Ph.D. Manchester, 1959) took over the office in 1975. Professor Thompson had come to the Institute of Optics in 1968 from the West Coast branch of Technical Operations, Inc. He had extensive experience in diffraction phenomena, holography, and image processing. Professor Thompson would be the dean for the next decade until he was appointed provost of the university in 1984. Professor Shapiro had been the associate dean for Graduate Studies since 1974, having replaced Professor Michael Hercher, Institute of Optics.

The department's undergraduate students during the years 1970–73 numbered around 100 each year, full-time Masters degree students between 10 and 12, part-time Masters degree students twice that, and Ph.D. degree candidates from 20 to 40. The number of undergraduate degrees for these same years fluctuated from 14 to 22 for the B.S., 8 to 15 for the M.S. and 3 to 9 for the Ph.D. Faculty publications were usually twenty or more per year, and sponsored research added up to about $300,000 per year.

Later in the 1970s, the number of electrical engineering undergraduates again grew both nationwide and at the University of Rochester. Students were aware of employment opportunities that existed for engineering graduates who could design very small integrated circuits for computers and communication devices. Consequently, their interests were in electronics, communication, and computer related courses. In 1976, Professor Kinnen began teaching the department's beginning courses in electronics, after Professor Cohen's adamant refusal to do so any more after having provided undergraduate instruction in electronics since the beginning of the department more than fifteen years before. The first semester course in electronics in the fall of 1976 had an enrollment that increased to 47 and then to 40 in the second semester. The required laboratories for these courses, however, had been designed for only eight workstations and 16 students. Large enrollments continued in these courses into the 1980s as the significance of Very Large Scale Integration (VLSI) became apparent for computer design and production and indeed for essentially all industries.

Many of the new courses introduced during these years illustrate not only the growth in number and breadth of the faculty, but also the emerging areas within the academic purview of electrical engineering. These courses along with some already mentioned include

| | |
|---|---|
| 1970 | Introduction to Solid-State Electronics |
| | Transistor Characteristics and Circuits |
| | Semiconductors, Transistors and Integrated Circuits |
| | Introduction to Computing |
| | Environmental Acoustics |
| | Digital Processing of Signals |
| 1971 | Introduction to Computer Engineering |
| | Introduction to Software Engineering |
| | Numerical Analysis |
| | Engineering Applications of Superconductivity |
| | Theory and Applications of Lasers |
| | Digital Processing of Signals |
| | Acoustical Physics |
| | Biomedical Instrumentation and Measurement |
| | Introduction to Biomedical Systems |
| | Bioelectric Phenomena |
| 1973 | Digital Circuits and Computers |
| | Digital Computer Programming Systems |
| | Digital Computer Organization and Design |
| | Microwaves |

| | |
|---|---|
| | Optimal Detection, Estimation and Digital Processing of Signals |
| | Nonlinear and Digital Control Systems |
| 1974 | Operating Systems |
| | Computational Methods for Engineers |
| 1975 | Electric Utility Engineering |
| 1976 | Bioultrasound |
| 1977 | Quantum Mechanics and Electronic Properties of Solids |
| | Geometric Modeling and Engineering Graphics |

Professor Kinnen introduced the course in Electric Utility Engineering in response to the 1973 oil embargo and interest nationwide in energy conservation. The course was meant to anticipate a need for graduating electrical engineers who would be familiar with electric power generation and distribution and with industrial power consumption. After three years, this course was taken over by David Fields (B.S. Rochester, 1973), an engineer with Rochester Gas and Electric Corporation, and then offered another three years. Instruction in this seemingly intrinsic part of electrical engineering could not be sustained in the department any longer, however. An increased job market for electrical engineers with some background in electrical power did not materialize and instructional resources had to be used for subjects preparing students in areas that did show increased job opportunities, such as in computers and solid-state electronics.

New members of the department were appointed largely to support existing research areas. However, none of this group, with the exception of Mr. Derefinko, established a long-term relationship.

- Edward Angel (Ph.D. USC, 1968), 1973; mathematical modeling of biomedical systems, dynamic programming, numerical analysis
- Neil C. Wilhelm (Ph.D. Stanford, 1973), 1973; stochastic models of computer systems, computer architecture and bus optimization
- David Pessel (Ph.D. UC Berkeley, 1974), 1974; computer engineering, ultrasonic fetal monitoring
- Charles V. Stancampiano (Ph.D. Rochester, 1975), 1975; nonequilibrium effects of superconductors
- David C. Farden (Ph.D. Colorado State, 1975), 1977; adaptive signal processing, convergence theory
- Arthur Frazo (Ph.D. Michigan, 1977), 1977; functional analysis, random processes, prediction theory

- B. Ross Barmish (Ph.D. Cornell, 1975), 1978; optimization theory
- Victor Derefinko (M.S.E.E. Virginia, 1967), 1979; instructor in EE, part-time

**Figure 22.** Dave Farden (R) and a student.

F. William Stephenson (Ph.D. Newcastle, 1965) spent the 1976–77 academic year as an R.T. French Professor on a unique faculty exchange program. He held an appointment as a senior lecturer in the Department of Electronic Engineering at the University of Hull in England. A major benefactor of Hull University was Reckitt-Coleman Ltd., the parent company of the R.T. French Company in Rochester. Largely unknown, an exchange program had existed since 1953 whereby someone from Hull would spend a year at the University of Rochester and the next year someone from Rochester would go to the University of Hull. The host company, either Reckitt-Coleman in England or R.T. French in the U.S., paid all expenses. In prior years, the exchanges had taken place only between faculties of the Colleges of Education. Professor Stephenson participated with the department's instruction in electronics and also completed a text while he was in Rochester.[3] Later on, he would join the faculty at Virginia Polytechnic Institute in Blacksburg, Virginia.

A number of joint appointments and part-time appointments were made during this period, in all cases augmenting research programs in biomedical engineering.

- James L. Cambier (Ph.D. Rochester, 1977), Department of Pathology, 1977–78
- Lawrence L. Chik (Ph.D. Rochester, 1969), Department of Obstetrics and Gynecology 1971–73
- Shirish Chikte, part-time in EE, 1979–80
- W. Bromley Clarke (Ph.D. Rochester, 1968), 1969–70
- E.R. Gordon Cook (Ph.D. MIT), part-time in EE, 1973
- Paul P. Lee (Ph.D. Rochester, 1977), Department of Radiology, 1978–81
- Vasant Saini (Ph.D. Rochester, 1979), 1979–85, Department of Obstetrics and Gynecology

- Leon Wheeless (Ph.D. Rochester, 1965), 1972–92, Department of Pathology

Visiting faculty and visiting scientists included:

- Goro Matsumoto, Hokkaido University, Japan, 1972
- Samuel Seraphim, Imperial College, England, 1971–72
- Frank Slaymaker, 1975
- F. William Stevenson, Hull University, England, 1976–77
- H.C. Torng, Cornell University, 1973–74

Other visitors, scientists and post-doctoral students:

- Gavin Armstrong, Leeds, 1977
- Shigeo Asami, 1970
- Gerald W. Bayer, Eastman Kodak, 1977
- John M. Boyse, General Motors Research, 1977
- Thomas F. Check, Calma, 1979–81
- Richard Clayton, 1977–78
- W. Burns Fisher, 1972–79
- Fred T. Goldstein, 1972
- Eugene Hartquist, 1974–86
- Peter Hoffmann, Hungarian Academy, 1977
- Robert A. Lee, Boeing Commercial Aircraft, 1977
- Mogens Levinsen, Copenhagen, 1976
- Konstantin K. Likharev, Moscow State University, 1976
- Richard J. Marisa, 1979–86
- Jeffrey Metzger, McDonald Douglas, 1979–81
- Alan E. Middleditch, 1972–73
- Thomas Nelson, Gleason, 1972–73
- Jan Ove Nilsson, 1978
- Hans W. Persson, 1977
- Uros Perusko, 1975
- Donald P. Peterson, Sandia, 1979–81
- Jan Ridder, 1977
- Nuriel Samuel, Xerox, 1974–76
- Ole Hoffmann Soerensen, Technical University of Denmark, 1974
- Richard Stark, NM State, 1978 and 1981
- Alexander Vystavkin, Institute of Radio Engineering and Electronics, Moscow, 1975
- Peter N.T. Wells, 1977
- Paul R. Zuckerman, 1972–74

Sabbatical leaves continued during the decade. Professor Titlebaum spent the 1970–71 academic year with the School of Public Health at Johns Hopkins University. At the same time, Professor Kinnen went to Hokkaido University and Tokyo University as a fellow of the Japan Society for the Promotion of Science and also to the University of Washington. Professor Shapiro journeyed to Denmark for his academic-year leave to be a visiting professor at Danmarks Tekniske Hojskole in Lyngby, Denmark. Professor Kinnen was also a ZWO Fellow at the Rijksuniversiteit Utrecht in the Netherlands for half a year in 1978 and again for the summer in 1981.

Professor Merriam's text, *Automated Design of Control Systems*, was published in 1974 followed by a monograph in 1978.[4] The next year, the third edition appeared of the *Handbook of Semiconductor Electronics*, edited by Professor Hunter.[5]

Looking back, it would appear that the faculty of the department responded notably to opportunities and challenges of the 1970s, though not without some members leaving. Professor Das resigned in 1973 to join the faculty at Rensselaer Polytechnic Institute in Troy, New York. Professor Wilhelm accepted a position with Hewlett-Packard in 1977, and Professor Angel transferred to the University of New Mexico in 1978. Also in 1978, Professor Flynn retired to professor emeritus status after completing nineteen years on the faculty. Professor Frazo left in 1980.

A change occurred in the U.S. patent system in 1979 by the passage of the Bayh-Dole Act. This would have an impact on the department in the decades to come as the Act granted permission to U.S. universities to license and profit from federally sponsored research.

**Figure 23.** Computer Studies Building.

# 5: The 1980s

> "Behind every able individual there are always other able individuals."
> *Chinese Proverb*

The third decade of the Department of Electrical Engineering can be characterized most dramatically by the impact of individuals. Some would be new to the department, some had long tenures, but each made an identifiable impression on the department as a whole.

The faculty of the department numbered twelve full-time professors and five part-time at the beginning of this decade:

- Professors Carstensen, Cohen, Hunter, Kinnen, Merriam, Shapiro and Voelcker
- Associate Professors Barmish, Titlebaum, Waag, and Leon Wheeless part-time
- Assistant Professors Farden and Stancampiano, and Shirish Chikte, Paul Lee, David Pessel and Vasant Saini part-time

Robert Sproull was president of the university, the title of chancellor having been dropped upon the retirement of W. Allen Wallis. Brian Thompson was dean of the College of Engineering and Applied Science. They held these positions until 1984. Associate deans were John C. Friedly, from Chemical Engineering, for Graduate Studies and Carlyle F. Whiting for Administration.

As in the 1970s, many individuals joined the faculty of the department during the 1980s and many left. The staff tripled in size. There was a major expansion of office and laboratory space, and a major investment was made in computing facilities. However, the country was in a period of recession in the early 1980s. Members of the faculty were under pressure to increase funded research activities, though at the same time, they were expected to spend more time with undergraduates and student laboratories. Engineering enrollments increased in an understandable reaction to the recession and better employment opportunities in engineering than in most other fields. But by the mid 1980s, the freshmen coming to the University of Rochester were again leaning toward social and health-related fields rather than the engineering disciplines. In response to this shift in student interests, a college-wide interdepartmental degree program was initiated to attract students with at least some interest in engineering but in less-structured, traditional curricula. Interdepartmental students were allowed, with some limitations, to create their own degree programs choosing courses in two or more departments of the college, the Institute of Optics, and the College of Arts and Science.

Professor Sidney Shapiro assumed the chairmanship of the department in September 1980, the position Professor Merriam had held since 1971. Professor Shapiro had been the associate dean for Graduate Studies for five years and was particularly active in creating options for both graduate and undergraduate students, such as the Special Opportunities Graduate Program for part-time graduate students. In January 1985, he would propose a Take Five Scholars program whereby selected students could elect to study outside their major field for a tuition-free fifth year. Although originally proposed only for undergraduates in the College of Engineering and Applied Science, the university's board of trustees approved the plan for all River Campus undergraduates. The first group of students to have their individual Take Five plans of study approved entered the program formally in September 1986.

**Figure 24.** Sidney Shapiro.

The Take Five Scholars program was originally suggested by Professor Shapiro during the time the Visiting Committee of the board of trustees was meeting with college administrators. As part of the discussions, two problems were cited relative to engineering student enrollments: 1) not enough freshmen were entering engineering, and 2) the requirements for an engineering degree left too little time for courses that would otherwise be part of a liberal education.

Professor Shapiro's suggestion addressed the second problem. It is interesting to note the steps that proved necessary for turning this idea into reality. Professor Shapiro sent a memo to the college deans and department chairs, with a copy to Provost Thompson and others, in which he outlined the concept and associated issues. Over the next several months, the idea was circulated around the River Campus colleges. The faculty met first in committees and then in their respective colleges to discuss the idea. In the College of Engineering and Applied Science, a meeting was held with students to learn their reaction to the plan. Finally, the plan was presented to the Committee on University Goals, which the new president, Dennis O'Brien, had established to consider changes in academic programs. Working over the summer, this committee recommended approval of the Take Five option for all River Campus undergraduates. The trustees met in January 1986 and approved the recommendation. Administrative procedures were set up quickly and students were first able to apply for the Take Five program in March 1986, fourteen months after Professor Shapiro first proposed the idea.

The success of the Take Five scholarship program is easily recognized. As of the end of 1999, over five hundred university undergraduates had taken advantage of this unusual opportunity. Topics of study by electrical engineering students who requested the program have included: Medieval Studies, Cultural and Universal Concepts of Beauty in Art and Architecture, Japanese Language and Culture, An Exploration of the Culture and Religion of Islam in the Middle East, African-American Experience Through Music and Literature, Filmmaking, German Studies, International Relations and the Swedish Language, Virtual Reality and the Understanding of Human Behavior, Photography and Film, and Australian Studies.*

Three academics began long associations with the department in 1981. Alexander Albicki (Ph.D. and Dr.Sc. Warsaw, 1973 and 1980) came in January. Thomas Y. Hsiang (Ph.D. UC Berkeley, 1977) and Kevin J. Parker (Ph.D. MIT, 1981) arrived in September.

Professor Albicki had been on the faculty of the Institute of Telecommunications, Warsaw Technical University, Poland, working with automata theory and digital telecommunication systems. However,

---

* Further details, including reports by students who have completed their programs, are available on the university's web site at: http://www.rochester.edu/College/CCAS/T5Abstracts.html.

the Solidarity movement had begun in Poland in 1980. As Professor Albicki was also a division deputy director, he was quickly recruited by the movement to represent the Institute. In this role he was highly visible as a member of Solidarity. When Professor Shapiro called him in October 1980 to invite him to Rochester as a visiting professor, he replied that he would come immediately. Immediately took less than three months; a visa was secured, flight plans made, and he arrived two weeks into the Spring semester. He was given the assignment of teaching Digital Logic Design to eighty-seven sophomores, taking over the class that Professor Voelcker had already started. The following year, 181 students signed on for the course, which had to be relocated to the lower Strong Auditorium. Professor Albicki and Logic Design became a duo that lasted for many years. His graduate students, meanwhile, would concentrate on the metastable operation of flip-flops and on timing and testing issues of microelectronic integrated circuits.

**Figure 25.** Alexander Albicki, left, with Menghui Zheng and his wife, Chaolin.

Professor Hsiang transferred from the Illinois Institute of Technology. His interests coming to Rochester were in the general areas of high-speed probing and noise in silicon devices. Gerard Mourou (Ph.D. Paris), a scientist at the L.L.E., had just developed an ultra-fast dye laser. Cooperating with Dr. Mourou initially, Professor Hsiang's research would lead to extraordinary studies of optoelectronic properties of devices and materials and non-equilibrium superconductor picosecond phenomena. Eventually Professor Hsiang would be able to reduce sampling times to the order of 200 femtoseconds.*

---

* Incidental to his research, Professor Hsiang's interest in the Japanese game of Go has led to several national and international competitions.

Interestingly, two of his students are carrying on his work as a married couple. Donald Butler received a B.S. degree from the University of Toronto and his Ph.D. degree in 1986. Zeynep Çelik graduated with honors from Bogaziçi University in Turkey and completed her Ph.D. degree in 1987. They were married in 1986, secured faculty positions in the Department of Electrical Engineering at Southern Methodist University, and since have established notable academic careers both in research and teaching.

**Figure 26.** Thomas Y. Hsiang.

Professor Parker's initial research would concentrate on the use of ultrasound to characterize tissue, thus complementing ongoing work by Professors Carstensen and Waag. Some of the older faculty members recall that Professor Parker, on the occasion of his visit to the department for an interview, gave a seminar that only lasted thirty-five minutes instead of the traditional fifty minutes. This left a positive impression on Professor Hunter who was known to have little enthusiasm for biomedical engineering.

**Figure 27.** Zeynep Çelik Butler and Donald Butler.

Roman Sobolewski (Ph.D. and Dr.Sc. Polish Academy of Science, Warsaw, 1983 and 1992) arrived as a research associate in 1980 to work with Professors Shapiro and Stancampiano on Josephson junction technology and high-speed pulse generation. Part of this work was characterizing variable-thickness Josephson junction bridges that were being fabricated at Cornell University. Professor Sobolewski would return to Poland and come back again in 1984 and 1987 to work with Professor Hsiang on high-speed optoelectronics and the propagation of picosecond signals on superconducting transmission lines. He would also study the dynamics of Josephson junction switching as an adjunct to Professor Stancampiano's work with picosecond photoresponses of lead and tin superconducting films. Gerard Mourou at the Laboratory for Laser Energetics was a collaborator in this research. When Professor Stancampiano left in 1983 to join the Eastman Kodak Research Laboratory in Rochester, Mark Bocko (Ph.D. Rochester, 1984) came over from the Department of Physics, in part also to study nonlinear effects of superconductors. However, as there was difficulty attracting research money to this area, efforts were redirected to low-temperature single-junction devices and quantum noise.

Subsequently, Alan Kadin (Ph.D. Harvard, 1979), Michael Wengler (Ph.D. Cal Tech, 1987) and Marc Feldman (Ph.D. UC Berkeley, 1975) joined the department and the superconductivity group to form one of

**Figure 28.** Roman Sobolewski.

**Figure 29.** Marc Feldman.

the largest academic programs in superconducting devices in the country. Prior to coming to Rochester, Professor Kadin had been investigating superconducting thin films and devices at Energy Conversion Devices, Inc., in Troy, Michigan. With the discovery of high-temperature superconductors, Professor Kadin redirected his research to problems of fabricating thin films of yttrium barium copper oxide (YBCO) and measuring the fast optical responses of these films. William Donaldson (Ph.D. Cornell, 1984) at the L.L.E. participated in these studies.

Professor Wengler was a Presidential Young Investigator, 1988–93, in recognition of his work with superconducting diodes for ultra sensitive detection at submillimeter wavelengths. Professor Feldman came from the University of Virginia where he had established a laboratory for superconducting device fabrication and applications of the devices to millimeter wave receivers. As a senior scientist, he would concentrate on superconductor-insulator-superconductor (SIS) devices and low noise receivers. He would play a key role subsequently in moving research efforts from predominantly single-junction superconducting devices to integrated circuits with many junctions.

In 1980 and 1981, microelectronic technology that later became known as very large scale integrated circuits (VLSI) was introduced to the academic community. VLSI was a radically new method for manufacturing solid-state electronic circuits. The finished circuits were essentially planar and had to be designed using a new computer-based procedure. Professor Kinnen promptly started a course on this design procedure but not without encouragement, advice, and help from Professor Albicki, some of the computer science faculty, and Professors Lynn Fuller and Roy S. Czernikowski, Rochester Institute of Technology. Little could be done during the first semester to carry out an actual design, as the design procedure at that time required a Xerox Altos computer with special software and a mouse. Two Altos machines, however, had just been delivered to the university, one to the Institute of Optics and one to Computer Science. Professor Albicki and graduate student Maciej Ciesielski had planned a simple logic gate design demonstration for the course. Using the ingenuity that comes naturally to graduate students, they arranged with some graduate students of the Institute of Optics for informal time late at night on the Institute's Altos machine. The arrangements included leaving a window of the building open for midnight access when all doors would be locked. The design demonstration was completed, and within a year the department had several Altos machines. VLSI has been part of the curriculum ever since.

In parallel with instruction in VLSI circuit design, Professor Hsiang introduced a companion course in VLSI fabrication. The course was changed later to VLSI Fabrication Principles. While simulation software was used for class instruction, a clean room was constructed in the Hopeman Building and was available for graduate student research by 1983.

Ari Requicha was given a faculty position in 1983, in addition to his ongoing appointments as senior scientist and associate director of the Production Automation Project. While continuing to develop a theory of solid-object modeling, the P.A.P. explored computer modeling systems for use with numerically controlled machines. A systems laboratory with a numerical controlled three-axis vertical-spindle milling machine was set up in the basement of the Hopeman Building, following some excavation around the building foundation necessary to install the milling machine. The department approved a graduate degree concentration in programmable automation based on courses in geometric and solid modeling, programmable machining systems, finite element analysis, and robotics. Gershon Kedem (Ph.D. Wisconsin, 1978) transferred his primary assistant professorship in computer science to electrical engineering but kept a secondary appointment in computer science. He also had a joint appointment with the L.L.E. Professor Kedem completed an initial design of a VLSI chip for an ingenious ray-casting computer for solid-modeling calculations. He left a year later for a faculty position at Duke University. This happened just before Renato Perucchio (Ph.D. Cornell, 1984), Department of Mechanical Engineering, joined the group. Over these years, a total of eleven industrial residents would spend from six months to two years as associates of the P.A.P.

Meanwhile, Professors Hunter and Cohen would retire to become professor emeriti, in 1981 and 1984 respectively. Professor Pessel resigned in 1981 to work at the L.L.E. and then followed Professor Lubin to the SOHIO Corporation in Cleveland. Robert L. McCrory, Jr. (Ph.D. MIT, 1973) replaced Professor Lubin as director of the L.L.E. in 1983. The next year Professor Barmish left for a faculty position at the University of Wisconsin.

The continued existence and growth of the L.L.E. after Professor Lubin left the university is a rare example of a research project that was started by one man and grew far beyond the individual. Good scientific people were attracted to the project and began working with Professor Lubin early on and financial support was found and sustained by his dogged perseverance. Eventually the L.L.E. took on a life of its own and by the end of the century had become a notable and visible part of the University of Rochester. In contrast, Professor Barmish's time with the department left little impact, except for two Ph.D. students. Professor Barmish had an intense interest in one aspect of optimization theory. He was known to deliver excellent lectures entirely without notes, although they consisted mostly of complex mathematical proofs and constructs. The applications of this theory were limited, however, and Professor Barmish chose to join a department with a large faculty.

At the close of the 1982–83 academic year, the department had fifteen full-time faculty and eleven sponsoring agencies for supporting various research projects. That year, there were twenty-five publications by the faculty and twelve conference proceedings, 37 full-time and 23 part-time graduate students, 86 entering freshmen, and 370 majors in electrical engineering. Fifty B.S. degrees were awarded. For added space, the sophomore circuits laboratory had to be moved to the top floor of Dewey Hall. The faculty voted on a one-semester teaching assistantship requirement for all Ph.D. candidates. Major research areas were: Biomedical Engineering and Bioultrasound, Computer Engineering and VLSI, Programmable Automation, Signal Processing and Communications, and Solid State Phenomena and Superconductivity. The next year, 1983–84, there would be nine sponsoring agencies supporting twenty research grants, thirty-seven published research papers and twelve reports in conference proceedings.

Undergraduate courses included subjects that were basic to a degree program in electrical engineering plus many others of particular interest to this faculty. These are listed by area.

---

I Computers and Computation
EE 101/102 Computing
EE 201/202 Computer Systems
EE 203/204 Computer Programming Systems
EE 206 Computational Methods for Engineers
EE 402 Stochastic Models for Computer Systems

II Systems and Control
EE 111 Circuits
EE 212/213 Systems and Control
EE 217/218 Electronic Systems
EE 410/411 Linear Systems/Nonlinear Systems

III Solid-State Electronics
EE 221/222 Solid-State Electronics
EE 327 Solid-State Electronics Laboratory

EE 420 Physics of Solid-State Devices
EE 424 Active Microwave Devices
EE 425 Superconductivity and the Josephson Effect
EE 429 Research Seminar in Solid-State Electronics

IV Fields and Waves
EE 231/232 Fields/Waves
EE 431 Microwaves
EE 433 Acoustic Waves

V Signals and Communications
EE 241/242 Signals/Communications
EE 440/441 Communication Theory
EE 446/447 Digital Signal/Image Processing

VI Special Interests
EE 253 Biomedical Systems
EE 256 Optimization with Applications to Large-Scale Systems
EE 258 Geometric Modeling and Engineering Graphics
EE 261 VLSI System Design
EE 416 Computer-Aided Design of Distributed-Parameter Systems
EE 437 Finite-Element Methods
EE 447 Programmable Machining Systems
EE 450 Bioelectric Phenomena
EE 451 Biomedical Ultrasound
EE 571 Advanced Geometric Modeling

---

Computers now available to the department included a VAX 11/780, DEC 2060, an IBM 4341, and a Cyber 175 (at the L.L.E.). John Lefor was hired by the department in 1984 to be responsible for purchasing and maintaining the many other smaller computers that were scattered around in the various offices and laboratories. He was also asked to build a local computer network, one of the first at the university to exist at a department level. By 1987 his responsibilities had increased to include other departments and he was given the title Manager of Technical Services and College Computing.

Basic studies of the bioeffects of pulsed ultrasound and nonlinearities in the propagation of ultrasound continued throughout the 1980s. Professor Flynn, now professor emeritus, was drawn back into active collaboration with the bioultrasound group. Professor Carstensen's laboratory had discovered that fruit fly larvae could be killed by pulsed ultrasound under conditions used in diagnostic ultrasound and when gas bodies were present in the respiratory system. It had been assumed by the medical ultrasound community that cavitation was highly unlikely with relatively short pulses. Professor Flynn's landmark report in 1982 corrected this erroneous perception. Charles Church, Department of

Biophysics, collaborated with Professor Flynn in a major theoretical study of cavitation resulting from short pulses of ultrasound.[1]

An electro-hydraulic lithotripter was introduced in the U.S. in the mid 1980s and quickly became the method of choice in the Department of Urology for the treatment of kidney stones. Professor Carstensen began studies of the side effects of the shockwaves generated by the lithotripter for stone comminution. This led in 1987 to a broadly based interdisciplinary collaboration which included Diane Dalecki, Ted Christopher, and Professor Parker, Department of Electrical Engineering; Sheryl Gracewski and Stephen Burns, Department of Mechanical Engineering; Asish Basu, Department of Earth and Environmental Sciences; Robert Mayer, Department of Urology; Nimish Vakil, Department of Gastroenterology; Eric Schenk and David Penney, Department of Pathology; Christopher Cox, Department of Biostatistics; Carr Everbach, Swarthmore; and David Blackstock, University of Texas at Austin. The lithotripsy study identified a number of new biological effects, many of which turned out to be relevant to diagnostic ultrasound as well.

Among the other biomedical engineering efforts taking place in the department, Professor Kinnen was co-director along with Martha Gram, Physical Therapist, Department of Pediatrics, of a project to design a lower-body orthosis to aid the mobility of children paralyzed from the waist down as a result of a birth defect called spina bifida. Also participating in the interdisciplinary effort were Franklin Peale, Orthopaedics, and Gregory Liptak, Department of Pediatrics; Gerald Tindall and James Brown, Rochester Orthopaedics, Inc.; and Marvin Gardner, Design Engineer. The orthosis, a parapodium, was demonstrated in a locally produced film entitled *All By Myself*, which had its premier at the Eastman House Dryden Theater on June 14, 1983. The film subsequently won six national and international awards, including the Golden Eagle Award of the U.S. Council on International Nontheatrical Events in Washington, DC. Ron Mix and Robert Bilheimer of Reel Images, Inc., Rochester, made the film. Forty-seven of the designed orthosis were fabricated and shipped around the country and Canada before production was turned over to the Variety Village Electrolimb Production Centre in Scarborough, Ontario, Canada.

The Rochester Center for Biomedical Ultrasound was created in 1986 with the encouragement of the department chair, Professor Shapiro. The mission of the center was to facilitate interaction among many activities in biomedical ultrasound that had developed

**Figure 30.** Rochester Parapodium.

independently throughout the university. Furthermore, it would increase the visibility of this work outside the university, in industrial and governmental laboratories and in other academic communities. Professor Carstensen became the founding director of the center. Charter members from the electrical engineering faculty included Professors Flynn, Mottley, Parker, Titlebaum, and Waag, and Professor Blackstock who was then a visiting professor in both the Electrical and Mechanical Engineering Departments. In total, early membership in the center was about thirty, a number that would increase to well over one hundred by the 1990s. Departments of the university that were represented included Anesthesiology, Biophysics, Biostatistics, Cardiology, Chemistry, Electrical Engineering, Family and Rehabilitative Medicine, Genetics, Mechanical Engineering, Obstetrics, Ophthalmology, Pathology, Pediatrics, Physics, Radiology, Surgery, and Urology. Faculty from clinical science at the Rochester Institute of Technology and from physical medicine and radiology at the Rochester General Hospital also belonged to the center. Professor Parker became director of the center in 1990 on the retirement of Professor Carstensen.

The department and the Rochester Center for Biomedical Ultrasound offered a series of summer short courses that were in reality state-of-the-art seminars. The first one was given in 1985 on acoustic cavitation. Speakers included Wesley Nyborg, University of Vermont, Larry Crum, University of Mississippi, and Professor Flynn. Shock Waves in Biomedical Ultrasound was offered next in 1987 with this list of topics:

- Physical origins of finite-amplitude effects
- Plane, spherically diverging and converging waves
- Waveform distortion and harmonic generation
- Shock formation and saturation
- Absorption of finite-amplitude waves
- Implication of nonlinear absorption for thermal surgery and hyperthermia
- Nonlinear properties of biological materials
- Nonlinear propagation in diagnostic ultrasound
- Imaging the nonlinear properties of tissue
- Biological effects of shock waves
- Cavitation in shock fields and lithotripsy

Charles Cain, University of Illinois, and Professors Blackstock and Parker participated.

The last of the series, in the summer of 1990, was on lithotripsy. Invited speakers included: Abraham Crockett, Department of Urology; Andrew Coleman, London; Michael Delius, Munich; Mark Adams and Robert Lerner (Ph.D., EE, Rochester 1977 and M.D., Rochester, 1978), Department of Radiology; Erdal Erturk, Department of Urology; and Roy Williams, Manchester. Here again, the work in biomedical engineering in the department is seen to affect the interests of many outside the department.

George Dennis O'Brien (Ph.D. Chicago, 1961) became the president of the University of Rochester in 1984, a position he would occupy for the next ten years. Brian Thompson moved from the deanship of the College of Engineering and Applied Science to the position of university provost. Miles Parker Givens (Ph.D. Cornell, 1942), Institute of Optics, became the acting dean for the next two years until the arrival of a new dean, Bruce W. Arden (Ph.D. Michigan, 1965). Professor Arden had

**Figure 31.** Bruce W. Arden.

been chair of the Department of Electrical Engineering and Computer Science at Princeton since 1973. His recent interests were in computer architecture, parallel computation, interconnection technology and multiprocessor systems. William E. Kiker (Ph.D. Tennessee) was now the associate dean for administration.

The space problem took on unusual proportions prior to the completion of a new building on the River Campus, the Computer Studies Building. The department had been occupying space in Dewey Hall and Gavett Hall in addition to about half of the rooms in the Hopeman Building. The move to the fourth and fifth floors of the new building in 1987 almost doubled office and laboratory space for Electrical Engineering. Thereafter, faculty, students, laboratories, and research areas of the department would be in just two locations, the Hopeman Engineering and the Computer Studies Buildings. The Department of Computer Science had the sixth and seventh floors of the new building. It was planned that the proximity of the two departments would encourage collaboration among individuals engaged in similar computer-related studies. This was seldom realized, however.

Coinciding with the move to the Computer Studies Building, the faculty completed an intensive review of the undergraduate curriculum. Subsequently, five upper-division courses would be required for all undergraduates: specifically, at least one course each in solid-state devices, analog electronics, signals, electric and magnetic fields, and waves. Students were also required to have three or four consistent sets of courses and a senior design project either in computer design, solid-state devices, communications systems, integrated electronics, or biomedical engineering.

The move to the new building was also the occasion for Professor Merriam to set up a new laboratory for computer design and computer architecture. Innovative design stations were constructed for student design teams rather than for individual students. The heart of each design station was a large capacity microprocessor that supported a variety of functions including Input/Output and communications with a student prototype micro-programmable processor. The microprocessor also allowed downloading of code, data, and diagnostic routines. A micro-crossassembler and other software tools resided on a host computer that was networked to each design station.

The five individuals who became members of the department faculty in the latter part of the 1980s would both contribute to existing areas of research and add new dimensions to the department character. Thomas B. Jones (Ph.D. MIT, 1970) joined the department in 1984 following three years at the Research Center of the Xerox Corporation in Webster, N.Y., and before that a faculty position at Colorado State University. His investigations of particle interaction with electromagnetic fields expanded in two directions. One direction was to study issues related to electrostatic safety. As a consequence, Professor Jones has been able, as a consultant, to recommend safety measures to minimize the risk of dust explosions in plastic processing plants and other manufacturing facilities that deal with powders. The other direction of his work has been to set up experiments on the dielectrophoretic manipulation of small particles in nonlinear fields. This work extended to modeling the conductive and dielectric properties of suspended particles and chains of particles, the latter including chains of biological cells.

**Figure 32.** Thomas B. Jones.

Vassilios D. Tourassis (Ph.D. Carnegie Mellon, 1985) came in 1985; and in 1986 Jack G. Mottley (Ph.D. Washington, St Louis, 1985), Robert Bowman (Ph.D. Utah, 1983) and A. Murat Tekalp (Ph.D. RPI, 1984).

Professor Tourassis's recent dissertation was a study of robotic controls, particularly on elbow manipulators. His arrival caused an administrative stir, however, as department attempts to get his J1 (international visitor) visa changed to an H1B (temporary specialty occupation worker) visa ran aground. His position on the faculty had been advertised with a Ph.D. degree requirement. Professor Tourassis had his thesis defense a month after the Carnegie Mellon University graduation in 1985 so his Ph.D. degree would not be awarded officially until the following June. The Immigration Office ruled that the advertised requirements were not satisfied and an H1 visa could not be issued. Eventually this was resolved, but in the process Professor Tourassis had to travel one day over the Peace Bridge from Buffalo to Canada and back again to satisfy one of the complex issues the Immigration Service was enforcing during this period.

Professor Mottley was attracted to the department as a result of his investigations at Washington University on the ultrasound characteristics of skeletal and cardiac muscular tissue. With Richard Moxley, Department of Neurology, he would extend his work to characterizing the skeletal tissues of patients with muscular dystrophy and to examining contrast agents for quantizing tissue perfusion from small vessels. Again the boundaries of biomedical engineering at the university were expanded by this interdepartmental study.

**Figure 33.** Murat Tekalp.

**Figure 34.** Victor Derefinko.

Professor Bowman had an appointment at the University of Vermont prior to relocating to Rochester. He received research support soon after he arrived from Analog Devices and the Siemens Corporation to assist his students working on the computer-aided design of analog circuits.

Professor Tekalp became a member of the department after three years in Rochester with the Eastman Kodak Company. His experience in image restoration and bispectrum estimation led to a number of digital image-processing projects. These included motion estimation and tracking, object-based coding, video compression, noise filtering, image indexing and super resolution. His student, Tanju Erdem (Ph.D. Rochester, 1990), was appointed an adjunct professor in the department while still an employee of Eastman Kodak. He would assist Professor Tekalp with graduate instruction in digital image processing. Some years later, Professor Tekalp would offer similar courses at Bilkent University in Ankara and Sibançi University in Istanbul, Turkey, during visits there and during a sabbatical year.

Another employee of the Eastman Kodak Company, Victor Derefinko, had an appointment with the department throughout the decade. After leaving Kodak, Mr. Derefinko spent several years at Johnson & Johnson. As a part-time lecturer, he provided instruction in electronics at different times for the department, the School of Medicine and Dentistry, and for non-electrical engineering students.

Others in the department were redirecting their interests as new research problems appeared in their areas of interest. Professor Albicki established a reputation for his work on the problematic issue of self-testability of VLSI systems. Professor Kinnen shifted his studies to applications of graph theory to problems of layout placement and routing. These studies of Professors Albicki and Kinnen were occurring at the same time other investigators around the country were also developing computer-aided design processes for the rapidly growing VLSI industry. Similarly, as the wireless communication industry was expanding in the country, Professor Titlebaum and his students were exploring frequency hop codes and developing theories of multiple access codes. Professor Farden was studying parameter estimation methods for time-varying system identification. Professor Waag's research now included the examination of both scattering and differential scattering of ultrasonic signals introduced for tissue characterization as well as studies of wave front distortion in weakly inhomogeneous media.

In still another of the department's research activities, Professor Parker, while working on sono-elastic imaging and image analysis relative to cancer treatment, was looking for a way to circumvent distracting image artifacts in the printouts from diagnostic equipment. Professor Parker and Theophano Mitsa, a graduate student, then discovered a half-toning technique they called a Blue Noise Mask. The Blue Noise Mask proved to be the fastest method known for producing high-quality half-tone images with ink jet plotters. The Mask and further developments were patented, although subsequent royalties that would come to the department would not be fully manifest until a decade later.

Professor Voelcker was on leave for one year, 1985–86, at the National Science Foundation in Washington, D.C. There he was the deputy director of the Design, Manufacturing and Computer Engineering Division, which he had helped found in 1985. Professor Voelcker resigned from the University of Rochester in 1986 to accept a chaired appointment in Mechanical Engineering at Cornell University. Following his departure, the Production Automation Project continued for a short time under Professor Requicha's direction until he accepted a position in the Computer Science Department at the

University of Southern California. Without the leadership of either Professor Voelcker or Professor Requicha, the P.A.P. disappeared at the University of Rochester. Professor Tourassis continued his robotic studies essentially in isolation. Also Professor Perucchio, Department of Mechanical Engineering, would continue his work on automatic mesh generation for the finite-element analysis of solid models.

Professor Saini left the university in 1985 to start his own company in Rochester, Advanced Computer Innovations. Professor Hsiang spent the 1987–1988 academic year at National Taiwan University as a visiting professor. And Professor Farden resigned in 1988 to join the faculty at North Dakota State University.

Visiting faculty in this department during the latter part of the decade were:

- Amalendu B. Bhattacharyya, IIT New Delhi, India, 1987–89
- David Blackstock, University of Texas at Austin, 1987–89
- Franco Bordoni, Un.de L'Aquila, Italy, 1989
- Enis Cetin, Bilkent University, Turkey, 1989
- Robert C. Chivers, University of Surrey, England, 1985
- E. Carr Everbach, Swarthmore College, 1989–90
- David J. Farber, University of Delaware, 1985
- Tomasz Kacprzak, Łódź Technical University, Poland, 1984–86
- Karan V. Kaler, University of Calgary, Canada, 1989–90
- Goro Matsumoto, Hokkaido University, Japan, 1986
- Peretz Meron, Israel, 1985–87
- Pier Gabriele Molari, University of Bologna, Italy, 1984
- S.K. Mullick, IIT Kanpur, India, 1986–87
- Andre Skovorada, Institute of Computer Science, Moscow, 1989

Other faculty appointments included:

- Mustafa A.G. Abushagur, Assistant Professor, part-time, 1984
- Marcelo Ang, Assistant Professor, 1988
- Victor Derefinko, Instructor, 1980–
- Marc J. Dumic, Lecturer part-time, 1982–83
- Christopher Hollot, Instructor, 1982
- James W. Horwitz, Adjunct, 1981
- James Minor, Instructor, part-time, 1984–88
- David Shields, Instructor, part-time, 1980–83
- Roman Sobolewski, Adjunct Associate Professor, 1987–90
- Ronald Yeaple, Adjunct, 1986–88

Visiting Scientists:

- Han Jigno, Shaanxi Microelectronics Research Institute, China, 1988–89
- Munawar Karim, St. John Fisher College, Rochester, 1989
- Tetsuo Kobyashi, Hokkaido Institute of Technology, 1987
- Neal Laurance, Ford Motor Company, 1985
- Peter Nauth, Germany, 1988–89
- Wesley Nyborg, University of Vermont, 1987
- Roberto Onofrio, Universita di Padova, Italy, summers 1989, 1990 and 1991
- Horace Thompson, Louisiana State University, 1987
- Frederick L. Thurstone, 1987
- De-xin Wu, Institute of Semiconductors, Chinese Academy of Science, 1985

Visitors and Research Associates:

- Silvia Ansaldi, 1984–85
- Del Armstrong, 1982–87
- Jeffrey P. Astheimer, 1981
- Brigita Blezurs, Leeds, 1980
- Graham Carey, Leeds, 1982
- Lulin Chen, 1985–88
- Charles Church, Illinois, 1986
- Alan L. Clark, Ford, 1982
- Charles L. DeRoller, 1982
- John Dodsworth, Leeds, 1981
- John Goldak, Carleton, 1983–84
- Mark R. Hopkins, 1980–82
- Christoph Hornung, T.U. Darmstadt, 1982
- James W. Horwitz, 1981
- Herve Huitric, Paris, 1981–82
- Peter Kane, Leeds, 1982
- Edward Maruggi, 1984–85
- Monique Nahas, Paris, 1981–82
- Dana S. Nau, Maryland, 1985–86
- David Parkins, 1983–85
- Anthony Saia, Leeds, 1982
- Jozef Sypniewski, 1984

Department outreach included participating on university and professional committees, providing editorial help on professional publications, and offering short courses on timely faculty research. The short courses included:

- Testing of Digital Circuits, Professor Albicki, 1987 and 1988
- Electromechanical Interactions of Dielectric and Magnetic Particles, Professor Jones, 1988 and 1989
- Robotics, Professor Tourassis, 1988
- Superconducting Electronics, Professor Wengler, 1988

Particularly of interest is the participation of department members in the activities of the Accreditation Board for Engineering and Technology (ABET). ABET has the task of evaluating and accrediting academic programs in departments of engineering and technology. This process includes periodic on-site visits to schools requesting accreditation. The Committee on Engineering Accreditation Activities (CEAA) of the Institute of Electrical and Electronic Engineers has the responsibility of recommending committee members to ABET for visits to electrical engineering programs. Three members of the department have served on this committee and on ABET teams: Professor Shapiro, 1984–92, Professor Waag, 1986–90, and Professor Jones, 1987–90. Collectively they made over a dozen evaluations of programs in electrical engineering and biomedical engineering.

Professor Carstensen was elected to the National Academy of Engineering in 1986. The next year, his monograph, *Biological Effects of Transmission Line Fields*, was published.[2] Professor Farden received the Edward Peck Curtis, Jr., Award for Excellence in Undergraduate Teaching in 1982. The citation that accompanied this award read "Anyone who writes on 'Statistical Design of Nonrecursive Digital Filters' and 'On Stochastic Approximation and Hierarchy of Adaptive Arrays' would need, almost of necessity, to be indulgent of and patient with student questions.... [Professor Farden] appears to be exemplary in this respect." Professor Tourassis received the 1989 Undergraduate Engineering Council Outstanding Teacher Award and Professor Waag was honored with a World Federation for Ultrasound in Medicine and Biology Best Paper Award for 1986.

Professor Shapiro stepped down as chair in 1989. He had guided the department through a total turnover and expansion of the technical and office staff, a doubling of the space occupied by the department, the comings and goings of seventeen members of the faculty, the graduation of some 433 B.S. degree candidates, and the granting of 127 M.S. and 18 Ph.D. degrees. Professor Kinnen then agreed to chair the department for the next three years until his retirement.

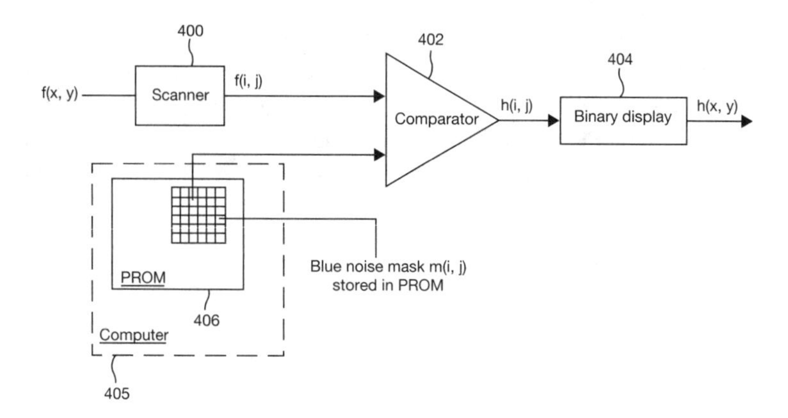


**Figure 35.1.** Illustration from Patent No. 5,708,518 issued to Kevin J. Parker and Theophano Mitsa, for the Blue Noise Mask.

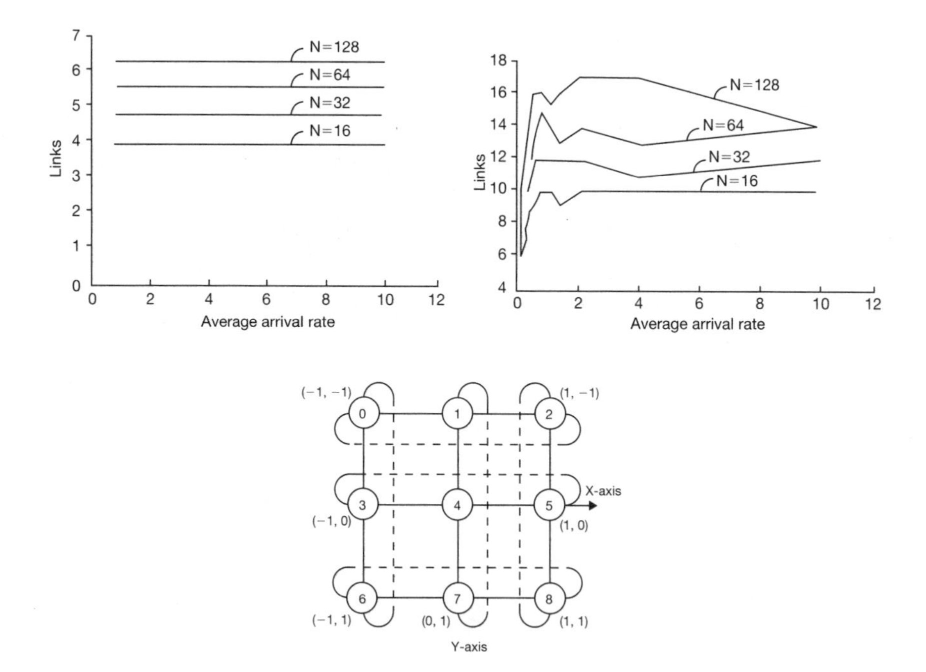


**Figure 35.2.** Illustration from Patent No. 5,218,676 issued to Mondher Ben-Ayed and Charles W. Merriam for a Dynamic Routing System for a Multimode Communications Network.

# 6: The 1990s

"We judge ourselves by what we feel capable of doing, while others judge us by what we have already done."
*Longfellow*

As in past years, new members joined the faculty of the department during the 1990s, others retired; new courses were proposed, the curriculum changed still another time, and the administration was shuffled. All this would be expected in a dynamic department. However, this decade saw an extraordinary increase in the breadth and extent of the department's research activities, the number of graduate students, the number of patents filed, and the number of visitors participating in the research efforts. The occurrence of these unusual events, nevertheless, was ironic. In 1990, the administrators of the university had decided to scrutinize the academic departments in an attempt to downsize or at least reduce costs at the department level. The timing of this attempt coincided, at least in electrical engineering, with renewed increases both in research opportunities and numbers of graduate student applications. The request to self-justify did not generate any enthusiasm within the department; rather interests were in taking advantage of new opportunities. As rumblings continued to portend downsizing, Professor Kevin Parker on his appointment as the department chair in 1992 saw the potential of extending the research and instructional programs in biomedical engineering and computer engineering, expanding upon the work that had been developing all along. The resulting expansion succeeded and the department's research initiatives attracted attention both within and beyond the university community. This chapter describes many of the changes and activities that took place up until the end of the century. Again, attention is often directed to individuals and the impact they have on events.

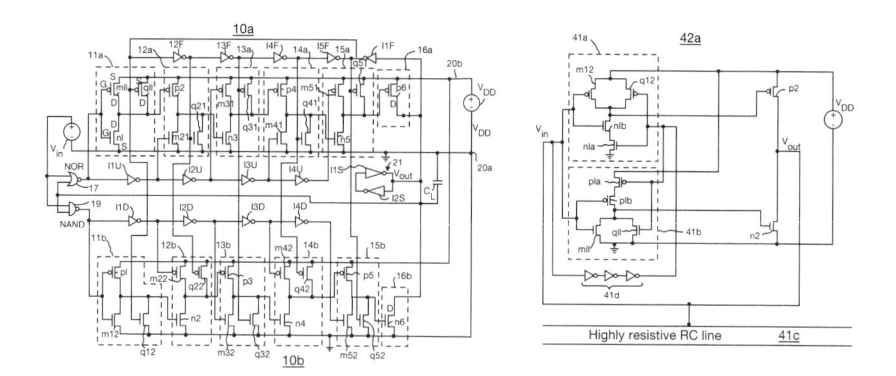


**Figure 35.3.** Illustration from Patent No. 6,163,174 issued to Eby Friedman and Radu M. Serareanu for Digital Buffer Circuits.

**Figure 36.** Edwin Kinnen.

For the record, the president of the university in 1990 was George Dennis O'Brien, Bruce Arden was the dean of the College of Engineering and Applied Science, and Edwin Kinnen the department chair. The faculty numbered seventeen with three additional as faculty associates. There were 210 undergraduate and 51 graduate students in the department, and research funds exceeded $6.5 million.

Professor Kinnen retired in 1992 as the chair and from the faculty to become professor emeritus and senior scientist. Professor Parker became the new chair in addition to continuing his responsibilities as director of the Rochester Center for Biomedical Ultrasound and a secondary appointment in the Department of Radiology. The next year, Thomas Humphrey Jackson (J.D. Yale, 1975), arrived as the ninth president of the university. Thomas LeBlanc (Ph.D. Wisconsin, 1982), Department of Computer Science, was the vice provost and dean of the faculty at that time. In a change that echoed the organization of the River Campus before 1940, the independent College of Engineering and Applied Science became the School of Engineering and Applied Sciences,

**Figure 37.** Kevin J. Parker.

returning to the university's College of Arts and Science. The latter was renamed the College of Arts, Sciences and Engineering. Bruce Arden then retired and Duncan Moore (Ph.D. Rochester, 1974), Institute of Optics, was appointed the dean of the new School. The official reason for the change from an independent college to a school within another college was given as financial—to take advantage of shared facilities. Eventually Dean Moore would accept an appointment as the associate director for technology of the White House Office of Science and Technology Policy and move to Washington, D.C. Professor Parker became the dean of the school in 1998 and Professor Fauchet the new chair of the department. Other administrative changes included Professor Mottley's appointment as associate dean for Undergraduate Studies in 1996, and Professor Titlebaum's appointment as vice provost for computing, a position he would enjoy, mostly, for four years.

Meanwhile, the college added a Bachelor of Arts in Engineering Science degree in yet another effort to accommodate engineering students with broad interests. The program, designed to be particularly attractive to students preparing for a career in manufacturing engineering or for graduate study in business, medicine, or law, included roughly one-third science and mathematics courses, one-third liberal arts courses and foreign languages, and one-third engineering courses. A Graduate Certificate Program was also introduced to attract local engineers into continuing education. After completing three courses in a specific area of research or technology, the student was awarded a certificate noting the area of advanced study. The pertinent areas in electrical engineering were Computer Design, Digital Signals and Image Processing, Medical Imaging, Superconducting Electronics, Telecommunications, and VLSI Digital Design.

Professor Shapiro organized a Senior Seminar on the ethical, social, economic, and safety issues that arise in the practice of engineering and transcend the technical. Students were introduced through readings, videos, discussions, and outside speakers to the complex issues of professional and personal responsibility that they could expect to confront after graduation and thereafter during a career in engineering. Professor Mottley took over as course coordinator when Professor Shapiro retired. Another new course, Communication Over Power Lines, Power Quality Issues and Nonintrusive Load Monitoring, was introduced in 1992 by Professor Albicki. And two years later Professor Bocko and David Headlam, Music Theory Department, Eastman School of Music, started formal instruction in yet another broadening of the field of electrical engineering, this time into the realm of recorded music. Their courses were titled Musical Sound: Science and Synthesis, and Introduction to Digital Music.

The department would appoint four to faculty positions while six others chose to retire or leave. Philippe Fauchet (Ph.D. Stanford, 1984) joined the department in 1990, coming from Princeton University with experience in femtosecond lasers, spectroscopy of semiconductors and optical characteristics of semiconductor material. These interests would broaden in Rochester to include porous silicon and nano-scale silicon for photonic and optoelectronic devices, and single electron transistors. Later, Professor Fauchet would become a senior scientist at the Laboratory for Laser Energetics and have joint appointments with the Institute of Optics, from 1994, and the Department of Physics, 1995–98.

The next year, Eby G. Friedman (Ph.D. UC Irvine, 1989) came east from the Hughes Aircraft Company in California to complement the department's instruction and ongoing research in VLSI. His studies of low power, gigahertz digital IC design, noise mitigation strategies, and clock tree topology synthesis quickly attracted support from government agencies as well as from local and West Coast companies.

**Figure 38.** Eby G. Friedman.

Professor Sobolewski returned in 1990 as senior scientist to continue his investigations of ultrafast phenomena in dielectrics, semiconductors, and superconductors. He would have secondary appointments with the L.L.E. and the Materials Science Program, and later be appointed professor of Electrical Engineering.

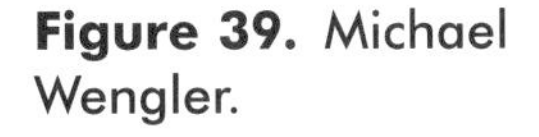

**Figure 39.** Michael Wengler.

**Figure 40.** Vassilios D. Tourassis.

Furthermore, in the comings and goings of the department, Professor Carstensen retired to emeritus professor in 1990 but continued for a number of years with an appointment of senior scientist. Professor Bowman left the university in 1992 to pursue a business venture, and later accepted a position at Rochester Institute of Technology. Professor Jerome Feldman this same year accepted a faculty position at the University of California at Berkeley. Professor Shapiro retired to emeritus status at the end of the fall semester 1993. Three years later, Professor Wengler decided to continue his career with the QualComm, Inc., in San Diego, California. Professor Tourassis resigned to return to Greece in 1998. Professor Tourassis's studies in robotics never gained much visibility within the university although his work with arm mechanisms might have provided an opportunity for interdisciplinary studies related to patient rehabilitation.

David Albonesi (Ph.D. Massachusetts, 1996) arrived in 1996, increasing faculty strength in computer engineering. He had been working on computer architecture for IBM and Prime Computers, Inc., for ten years before returning for a Ph.D. degree. His dissertation was on complexity issues of adaptive computing processes. In cooperation with Professor Sandhya Dwarkadas, Department of Computer Science, his research would soon extend to adaptive chip architecture and computer architectural optimization methodologies.

**Figure 41.** David Albonesi.

In 1991 and 1992, professional peers honored Professor Carstensen. A festschrift was held at the university, following an old German custom of highlighting a professor's career by a gathering of former students and colleagues. Papers presented at the two-day conference were published the following year in a special issue of the *Journal of Ultrasound in Medicine.* Professor Carstensen was also recognized in 1992 with a Career Achievement Award from the Society of Engineers in Medicine and Biology. The award, in recognition of his work on the effects of electric and magnetic fields and ultrasonic energy on living organisms, was presented at a ceremony in Paris, France. Professor Blackstock was elected to the National Academy of Engineering in 1992, thus joining Professor Carstensen with this distinction.

A number of sabbatical leaves were granted.

- Professor Jones to the Fukui Institute of Technology, Fukui, Japan, and the University of Calgary, Canada, 1990
- Professor Kadin to the University of Minnesota, 1995
- Professor Tekalp to Bilkent University, Turkey, 1992
- Professor Titlebaum to the University of Rhode Island, 1991
- Professor Wengler to the Hypres Corporation, 1993

Also in 1990 Professor Waag became a fellow of the Alexander von Humboldt Foundation at the Ruhr-Universitat Bochum, Bochum, Germany. The next year he was visiting professor at the University of Paris VII, and the Tokyo Institute of Technology. Professor Hsiang was on leave during 1991–92 for a year at the National Science Foundation in Washington, D.C., as the program director for the Solid-State and Microstructures Program. Professor Kinnen traveled to the Indian Institute of Technology, New Delhi, India, under a U.S.-India Scientist Exchange. He also spent four weeks in Albania in the fall of 1992 as a consultant to the Department of Electrical Engineering at the Polytechnic University of Tirana. This was arranged through the International Executive Service Corps in Stamford, Connecticut.

Research in the department was noticeably productive. In 1992, the university received a Research Initiative Grant on Low Temperature Superconducting Digital Electronics for continuing research in superconductivity. The Superconductivity Group had changed its name to the Superconducting and Quantum Electronics Group as they began to investigate combinations of single junction superconducting devices for digital integrated circuits. The proposal submitted for this grant had made a point of describing relevant research capabilities in the department and listed a total of nine faculty members as investigators. The investigators included those who had been working directly with single junction devices and high-speed applications, Professors Bocko, Friedman, Hsiang, Kadin, Shapiro, and Sobolewski. The three others from the department that were listed—Professors Albicki and Feldman and Visiting Professor Andrzej Krasniewski—had experience in digital integrated circuits, computer logic, and architecture. Professor Krasniewski had returned to Rochester in 1991 after an absence of six years.

The research under this University Research Initiative Grant progressed from single-device circuits to multiple-device circuits. The culmination of this progression in the late 1990s was a high-speed digital filter constructed of about four thousand Josephson junctions. The filter operated at a 20 GHz clock frequency using a special resistive single-flux quantum, or RSFQ, logic developed for two-terminal switching devices. Fabrication took place at the Science and Technology Center of the Westinghouse Corporation in Pittsburgh, PA. With similar research programs at the State University of New York at Stony Brook and the University of California at Berkeley, the Rochester group decided to concentrate on developing computer software to simulate these circuits and on determining the relevant process parameter variations needed

**Figure 42.** Mark Bocko, right, with graduate student Jonathan Habif, ready to test a 10 GHz RSFQ ring oscillator clock, 2000.

**Figure 43.** Alan Kadin, right, and graduate student Dorek Mallorg.

for simulation. Research also extended to studies of a quantum gate using a Josephson junction as the basic element of an integrated computer circuit but with an added advantage of having built-in test circuits. By the end of the century, it is estimated that members of the Department of Electrical Engineering at the University of Rochester had supervised the greatest number of Ph.D. dissertations anywhere in the field of superconductivity. In 2000, Professor Kadin decided to return to industry, joining the Hypres Corporation in Elmsford, N.Y.

**Figure 44.** Diane Dalecki.

In the 1990s, Professor Carstensen's bioultrasound laboratory, now under the leadership of Diane Dalecki (Ph.D. Rochester, 1993), expanded its bioeffects research. This resulted in showing that any tissue containing gas bodies or a microbubble contrast agent was particularly susceptible to damage from exposure to pulsed ultrasound levels widely used for medical diagnosis. Studies of the bioeffects of mid-kilohertz frequencies were also initiated, as it had been shown that these were the dominant frequencies in the spark-generated shockwaves used in lithotripsy. As a result of these studies, Professor Dalecki was able to contribute to the U.S. Navy's bioeffects database on exposure guidelines for divers in the vicinity of high-power, low-frequency sonar sources. Other studies, in cooperation with Charles Francis, Department of Vascular Medicine, demonstrated a nonthermal mechanism for ultrasound enhancement of the lyses or breaking up of blood clots.

Professors Parker and Tekalp with Ruola Ning, Department of Radiology, were examining the restoration, enhancement, and analysis of three- and four-dimension magnetic resonance images, MRI. They were applying some of Professor Tekalp's research on digital image processing while extending Professor Parker's recent investigations into three- and four-dimensional sonoelasticity. Ted Christopher (Paul E.) in the late 1990s added another facet to the group's work in bioultrasound. He discovered that the second harmonic in the reflected signal of a typical ultrasound system due to nonlinear distortion could be used to generate dramatically sharper and clearer images. This idea was patented and the technology is now widely used in medical ultrasound. Somewhat tangentially, Professor Waag was refining ultrasonic image processing by using the reflected waves that are scattered by the imaging object.

Professor Parker, with Xucai Chen and Karl Q. Schwarz, Department of Cardiology, would improve blood flow visualization with harmonic imaging, that is, by recording the harmonics and subharmonics of the ultrasonic signal reflected from microbubbles in the flow stream. They also demonstrated the advantages of processing the sum and difference frequencies of a dual-frequency ultrasonic tone burst. In another innovative use of ultrasound, Professor Mottley attached an ultrasonic sensor to the end of an oxygen delivery tube. This created an inexpensive endotracheal tube for comatose patients for use, for example, by emergency personnel to detect an incorrectly inserted breathing tube.

In recognition of his international reputation for studies in acoustical phenomena, Professor Waag became the Arthur Gould Yates Professor of Engineering in 1994. At this time, his bioacoustics research had extended to refining high resolution imaging using aberration correction and quantitative imaging using inverse scattering. The Arthur Gould Yates Professorship was established in 1927 under a $100,000 endowment given by Mrs. Yates in memory of her husband, who had

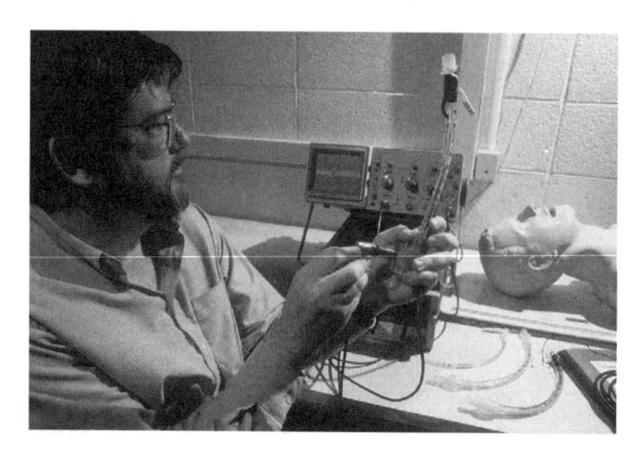

**Figure 45.** Jack Mottley preparing an endotracheal tube.

been a trustee in the 1890s. Professor Waag was only the fifth member of the faculty to hold this endowed chair.*

Professor Jones began working on the use of electric fields to manipulate small liquid masses on a substrate and to dispense sub-nanoliter droplets. This work has potential implications for a laboratory on a chip. Professor Hsiang, who was also a senior scientist at the L.L.E., demonstrated ultra-fast silicon photo-detectors operating at speeds of 75 gigahertz and beyond. He then used these detectors as sampling devices to characterize other ultra-fast devices such as the High Electron Mobility Transistor (HEMT), and the Resonant Tunneling Diode (RTD). Professor Titlebaum and his students were studying congruential coding for code-division multiple-access, particularly for use with cellular telephone systems. Professor Sobolewski and a colleague from the Moscow State Pedagogical University developed a hot-electron photo detector, a superconducting device for single-photon detection operating in the infrared spectrum.

Professor Albicki received support from a consortium of electric utility companies to study selected coding schemes for use in power line communications. Later, the support was extended to work on a problem related to the now widespread interest in energy conservation. This was the problem of measuring nonintrusively the energy consumption and switching patterns in residential, commercial, and hospital buildings. At about the same time, he offered a course that considered the background of these and other timely issues of the public power utilities. Robert Jones and David Shields, engineers with the Rochester Gas and Electric Corporation and graduates of the

---

* The others were Professor Gavett, 1927–42, Professor Belknap, 1946–48, Professor Martin Lessen, Department of Mechanical Engineering, 1967–83, and Professor Carstensen, 1988–90.

department, participated in this course. Professor Merriam and his student, Mondher Ben-Ayed, received a U.S. patent entitled "Dynamic Routing System for a Multimode Communication Network." Professor Tekalp extended his digital image processing work to new video signal processes such as image/video segmentation, fast restoration, and secure access. Visiting Professor Krzysztof Gaj offered a Special Topics course, Cryptology and Computer-Network Security, during the summers of 1995, 1996 and 1997. The course considered such topics as privacy and security on computer networks and financial fraud and theft using the Internet.

Twenty patents were issued to the university as a result of research carried out in the department during the 1990s. A few of these, such as the Blue Noise Mask patent by Dean Parker and his student Theophano Mitsa, proved to be particularly remunerative. The blue noise mask technology was licensed for ink-jet printing by about a dozen companies, including the Seiko Epson Corp. and the Hewlett-Packard Company. In a survey of publications in seventy journals of electrical engineering published 1986–90 and indexed by the Institute of Scientific Information, the University of Rochester ranked second only to Stanford University in the average number of citations per paper in a list of electrical engineering departments, with at least fifty publications during this period.[1] Not bad for a small department.

**Figure 46.** Theophano Mitsa, right, working with Professor Titlebaum on a student project.

Some of the courses that appeared in the department bulletin at this time further characterize new faculty interests within the department.

- Introduction to Digital Music
- Musical Sound: Science and Synthesis
- Introduction to Opto-electronics
- Electric Power Quality

- Fabrication Principles in Solid-State Electronics
- Superconducting Electronics
- Electromechanics of Particles
- Heterogeneous Dielectrics
- Digital Imaging Technology, Pattern Recognition
- Digital Video Processing
- Medical Imaging
- Device Characterization for IC Design
- Performance Issues in VLSI/IC

Special Topics courses were offered in:

- Issues in Manufacturing
- Reconfigurable Microprocessor Design
- Data and Image Coding, and
- Science and Technology for Future Computation

**Figure 47.** Philippe Fauchet.

Dean Moore was granted leave in the middle of the 1997–98 academic year to accept an appointment in Washington, D.C. Professor Parker then became dean of the School of Engineering and Applied Science and Professor Fauchet assumed the chair of the department. In 1998, the Department of Electrical Engineering changed its name to the Department of Electrical and Computer Engineering. This was done to better reflect the continually increasing activities of the department in many aspects of computer engineering, from elemental circuitry and logic to computer architecture and diverse applications. A year later, curriculum modifications would be made to further integrate precepts shared both by electrical engineering and computer engineering. These modifications also included a new freshman level course in Signals and Circuits, and a two-semester capstone design sequence for seniors. The latter was organized around the concept of design teams, in contrast to older design courses that emphasized individual effort.

Other initiatives appeared around the university during the late 1990s that would precipitate a major change in the academic organization of biomedical engineering on the campus. The university's Renaissance Plan had been implemented in 1995; this reduced and refocused undergraduate courses on the River Campus. The medical center at the university had launched a major strategic plan that included both the construction of a new research facility and the recruitment of a number of senior faculty to enhance the university's reputation as a major center for biomedical research. At the same time, the School of Medicine and Dentistry and the School of Engineering and Applied Sciences were promoting faculty interactions to further strengthen research and course instruction in biomedical engineering. Dean Moore and Richard Waugh (Ph.D. Duke, 1977), Department of Pharmacology and Physiology, who had a secondary appointment in the Department of Mechanical Engineering, had received a development grant from the Whitaker Foundation in 1996 for the purpose of reinstating graduate degrees in biomedical engineering at the University of Rochester. These degrees would be granted by way of a new interdisciplinary program in biomedical engineering that would admit students directly and include faculty from both the School of Engineering and Applied Sciences and the School of Medicine and Dentistry. While graduate degrees in biomedical engineering had been authorized in the College of Engineering and Applied Science, none had been granted for some time, prompting Dean Arden to terminate the degrees. The new interdepartmental program had twenty core faculty, some from each of the engineering departments on campus, others from various medical departments. Salaries for three faculty members were provided by the grant. Professor Waugh became the chair of the program.

A second Whitaker development grant was received two years later to help establish an undergraduate degree program in biomedical engineering to complement the graduate degree program then in existence. This B.S. degree was approved and the first graduates received their degrees in 1999. The timing seemed right at this point for then Dean Parker and Professor Waugh to propose an independent and interdisciplinary Department of Biomedical Engineering at the university. The board of trustees approved the department in January 2000 and Professor Waugh was again selected to be the chair.

The decision to create a separate Department of Biomedical Engineering ended the debate that started back in the 1960s over the merits of formally separating biomedical engineering from an existing engineering department. As mentioned earlier, the arguments were about potentially unrealistic career expectations and the greater value of a well-known engineering degree to someone entering a cross-disciplinary field. However, almost forty years later, biomedical engineering had become a visible profession in schools of engineering, in most medical research facilities and in many health-related industries. So the decision to have a Department of Biomedical Engineering at the University of Rochester appeared to be a timely one. Further evidence that this was an appropriate decision was the amount of faculty support. The new department, in addition to eight professorial appointments, included twenty-four joint or secondary appointments from fifteen departments around the university and eleven adjunct appointments from nine departments. Professors Fauchet, Mottley, and Parker had secondary appointments from electrical engineering.

Continuing the practice in the department of offering short courses, one on the Nonlinear Revolution in Medical Ultrasound was held in June 1998. Organized by Professor Parker and Deborah Rubens, Department of Radiology, the speakers included six other members of the Rochester Center for Biomedical Ultrasound and also Professor Blackstock. This came approximately thirty years after the seminal papers on nonlinear ultrasound that Professor Blackstock had published while he was a member of the department faculty.

Clearly, the Department of Electrical and Computer Engineering had a major role in establishing the Department of Biomedical Engineering. Professor Healy's legacy, intended or not, had provided a multitude of research reports, seminars, courses, and graduates in biomedical engineering. These all followed from a wide range of investigations, from electrical, magnetic, and acoustic field effects in tissues to contributions made to medical diagnoses using ultrasound and innovative imaging methods. The department provided office and laboratory space for new faculty and staff. Martin Anderson (Ph.D. Duke, 1997) came into the department initially until formalities for the new department were in place. Professor Dalecki, now an assistant professor, transferred over from electrical engineering to biomedical engineering as did Sally Z. Child, concluding a notable thirty-five-year career in electrical engineering.

The 1990s also saw the creation of the Center for Electronic Imaging Systems. The objective of the center was to establish a visible entity and working space for a number of collaborative imaging investigations. These investigations had existed within and between the

Departments of Electrical Engineering, Computer Science, and the Institute of Optics. Other collaborations were ongoing between members of the faculty of the University of Rochester and the faculty of the Rochester Institute of Technology, as well as with individuals in the Xerox Corporation and the Eastman Kodak Company. As expected, Professors Parker and Tekalp were part of the organizational effort of the center. Professor Nicholas George, Institute of Optics, was the director from 1993 to 1998. Michael Allen Kriss (Ph.D. UCLA, 1969), who had retired from the Eastman Kodak Company in 1994, was appointed senior scientist in electrical and computer engineering and the associate director of the center. In turn, a cross-disciplinary curriculum was established for imaging science and for electronic imaging systems. Eventually, twenty-four individuals would participate in the center, coming from the Department of Electrical and Computer Engineering, the Department of Computer Science, and the Institute of Optics along with scientists from the Rochester Institute of Technology and the collaborating corporations. Professor Friedman assumed the directorship of the center in 1999.

Professor Fauchet and Alice Pentland, Department of Dermatology, established the second new center with objectives somewhat orthogonal to mainstream biomedical research. This new direction was also a marked change from Professor Fauchet's established research programs in semiconductor materials and high-speed phenomena. Named the Center for Future Health, it had an innovative mission starting in 1999 to create devices and processes that could be used by individuals in their homes to monitor their state-of-health or a chronic condition. The objective was to make technology affordable and easy to use, as the future of health care was expected to be more dependent on health maintenance and self-monitoring. Alexander Pentland, academic head of the MIT Media Lab, was appointed an external director of the center. Associates included faculty from the Departments of Chemistry, Computer Science, Religion, and Psychiatry, the latter to help to understand the social implications of the center's research. Investigators at the Georgia Institute of Technology and the University of Toronto were also participants. At its creation, the center received a grant from the W.M. Keck Foundation and support from Eastman Kodak Company and the Intel Corporation.

John Lefor resigned in 1990 as manager of computing resources to join the Microsoft Corporation in Redmond, Washington, where greater opportunities beckoned. Dikran Kassabian took over this management function for the School of Engineering and Applied Sciences. John Simonson then replaced him in 1995 when Mr. Kassabian left for the University of Pennsylvania.

In 1995 Cambridge University Press published Professor Jones's monograph in which he generalized the theory of particles in non-uniform electric and magnetic fields.[2] The same year, Professor Tekalp's book was published; this described digital image and video processes for multimedia and the technology underlying high definition television.[3] Kluwer Academic Publishers issued two volumes of technical material written for IEEE publications and edited by Professor Friedman.[4] Then in 1999, Professor Kadin's book, *Introduction to Superconducting Circuits*, appeared. This text was unique in describing both the physics of superconductors and their applications from a circuit perspective.[5] Professor Shapiro received the Undergraduate Engineering Council Outstanding Teacher Award in 1993. Professor Hsiang won the U.S. Go Open Championship in 1990 and 1997, and the North American Go Invitational in 1995. He was also the U.S. representative in the Go World Championship competition in 1998. Sponsored research for the department averaged $2,701,500 for the years 1998–99.

Visiting faculty appointments during the decade:

- David Blackstock, University of Texas at Austin, 1990–93, 1996–99
- Krzysztof Gaj, Warsaw Technical University, 1995–98
- Andrzej Krasniewski, Warsaw Technical University, 1990–97
- Witold Kula, Polish Academy of Science, Poland, 1995–96
- Makoto Tabei, Tokyo Institute of Technology, Japan, 1997
- Norimichi Tsumura, Chiba University, Japan, 1999
- Leonid V. Tsybeskov, Mechnikov Odessa University, 1993–99
- Margit Zacharias, Germany, 1996–97

Visiting scientists included:

- Chang Wen Chen (Ph.D. Illinois, 1992), 1995–96
- Xucai Chen (Ph.D. Yale, 1991), 1994
- Grigori Goltsman, Moscow State Pedagogical University, Russia, 1997–99
- Zaegyoo Hah, Kong Ju National University, Korea, 1999
- Frank Hegmann, Canada, 1994
- Konstantin Iline, Russia, 1998
- Sven Gunnar Mikael Lindgren, Sweden, 1996–97

- Andrey Lipatov, Russia, 1999
- Dong-Lai Liu, China, 1996–98
- D. Tetsuya Nagata, Hitachi Ltd., Japan
- Alexey Semenov, Russia, 1999
- Larry Shi, China, 1993–94
- Wolfgang Lang, University of Vienna, Austria, 1997
- Jury Vandyshev, Russia, 1993–96
- Masao Washizu, Seikei University, Japan
- Wen-Sheng Zeng, China, 1996–97

There were several secondary appointments of faculty from other departments:

- Sandhya Dwarkadas, Computer Science, 1998–99
- Professor Nicholas George (Ph.D. Caltech, 1959), Institute of Optics, 1993–99
- Stephen Fred Levinson (M.D. and Ph.D. Purdue, 1981), Department of Physical Medicine and Rehabilitation, 1993
- Professor Richard Mandelbaum (Ph.D. Princeton, 1970), Department of Mathematics, 1991–94
- Ruola Ning (Ph.D. Utah, 1989), Department of Radiology, 1998
- Denham S. Ward (M.D. and Ph.D. UCLA, 1975), Department of Anesthesiology

Of note, Professors George, Levinson, and Ward each received a Ph.D. degree from a department of electrical engineering.

Adjunct faculty appointments were:

- Victor Derefinko (M.S.E.E. Virginia)
- A. Tanju Erdem (Ph.D. Rochester, 1990), 1996–99
- E. Carr Everbach (Ph.D. Yale)
- Michael A. Kriss (Ph.D. UCLA, 1969), 1996–99
- Alexander Pentland (Ph.D. MIT, 1982), 1997–99
- Eli Saber (Ph.D. Rochester, 1996), 1997–99
- M. Ibrahim Sezan (Ph.D. RPI), 1992–98

Lecturers were:

- Paul E. Christopher (Ph.D. Rochester, 1993), 1995–98
- Julie Petronio, 1995
- Kendall Stephenson (Ph.D. Rochester, 1994)
- Jeffrey Wilcox, 1996–98

Visiting scholars and research associates included:

- Roman Adam, Slav Republic, 1995
- Julien Bailat, Switzerland, 1999
- Gozde Bozdagi, Turkey, 1994–95
- Isil Celasum, Turkey, 1997
- Chang Wen Chen (Ph.D. Illinois, 1992), 1993–97
- Lulin Chen, China, 1995–98
- Robert M. Cramblitt, 1995–97
- Carsten Draeger, Germany, 1998–99
- Ian Egerton, England, 1996–97
- Erwan Fradet, Switzerland, 1998–99
- Krzysztof Gaj, Poland, 1994–95
- Wolfgang Gob, Austria, 1994
- Ting Gong, China, 1990–94
- Bo Ran Guan, China, 1993–95
- Bilge Gunsel, Turkey, 1995–98
- Erling Hoie, Norway, 1996–97
- Fulco Houkes, Switzerland, 1998–99
- Sung Huang, Taiwan, 1990
- Marc Jungo, Switzerland, 1998–99
- Tomas Jansson, Sweden, 1995–96
- Mark Johnson, 1996–97
- Man Bae Kim, Korea, 1996
- Hideki Koyama, Japan, 1998–99
- Witold Kula, Poland, 1992–95
- James Lacefield, 1999
- Eduardo Lugo, Mexico, 1996–99
- Dong-Lai Lui, China, 1992–95
- Svet Maric, Serbia, 1995
- T. Douglas Mast, 1993–96
- Sushrut Mehta, 1990–92
- Laurent Montes, France, 1998–2000
- Daniel Phillips, 1998
- Tasso Sales, Brazil, 1998–2000
- Bozenna Sobolewska, Poland, 1993
- Laurent Souriau, France, 1996–98
- Yutaka Urino, Japan, 1996
- Petrus Van Beek, Netherlands, 1996–98
- Igor Vernik, Russia, 1998–99

- Jens von Bahren, Germany, 1994–97
- Leszek Wronski, Poland, 1990–92
- Daniel Young, 1990–91

It is worth recalling, as this history concludes, that members of the Department of Electrical and Computer Engineering had been active participants in creating and maintaining collaborative studies with numerous departments and disciplines throughout the university and elsewhere. These activities grew into six research centers of the university by 2000.

1. The Rochester Center for Biomedical Ultrasound with over one hundred members, six from the Department of Electrical and Computer Engineering. Others came from the Departments of: Anesthesiology, Biomedical Engineering, Biochemistry and Biophysics, Biostatistics, Cardiology, Earth and Environmental Sciences, Gastroenterology, Laboratory for Animal Medicine, Mechanical Engineering, Neurosurgery, Obstetrics and Gynecology, Ophthalmology, Pathology and Laboratory Medicine, Physical Medicine and Rehabilitation, Radiation Oncology, Radiology, Surgery, Urology, and Vascular Medicine. Physicians from the Rochester General Hospital, faculty members from the Rochester Institute of Technology, and five visiting scientists were also associated with the center. This was the largest known collaboration anywhere of ultrasound researchers in one location.
2. The Center for Electronic Imaging Systems with thirty faculty members, eight from the Department of Electrical and Computer Engineering, and others from the Institute of Optics, the William Simon Graduate School for Business Administration, the Laboratory for Laser Energetics, the Departments of Computer Science and Radiology, and the Rochester Institute of Technology. The center also had an association with twenty-eight companies from around the United States, including the Eastman Kodak Company, the Xerox Corporation, the Bausch and Lomb Corporation, the Siemens Corporation, the Corning Glass Company, Sony, the Boeing Corporation, IBM and Intel.
3. The Center for Visual Science, with a long history that included Professor Cohen in the 1950s and 1960s. In 1999, twenty-five faculty members were part of the center but none from the department. Cooperating departments included: Computer Science,

Neurobiology and Anatomy, Neurology, and Ophthalmology, plus the Institute of Optics. The center also had working relations with the Bausch and Lomb Corporation and the Lawrence Livermore Laboratories in California.

4. The Center for Future Health with four from the department, plus individuals from the Departments of Biomedical Engineering, Chemistry, Computer Science, Dermatology, Emergency, Family Medicine, Neurology, Physical Medicine and Rehabilitation, and Psychiatry. The Monroe Community Hospital and the Media Laboratory at MIT were also represented in the center.
5. The Center for Optoelectronics and Imaging, established as a facility for collaborative imaging research among universities and local imaging companies. Two members of the department were working with this facility.
6. The Center for Superconducting Digital Electronics was created in 1993. Six members of the department have been included along with industrial associates, the Hypres Corporation, TRW, Inc., and the MIT Lincoln Laboratory.

Professor Brian Thompson, Institute of Optics, in his inaugural address as university provost on May 28, 1975, stated:

> We must continue to take advantage of the unique situation at Rochester that easily allows for interdisciplinary research. This interacting effort will continue to be stimulated with various parts of the River Campus as well as other colleges of this University. These activities are particularly important since it allows the engineer to use his skills to solve a sophisticated and challenging problem in a unique and original way.... I believe we can...justifiably claim that we have one of the best colleges of Engineering and Applied Science of its size in this country.

Twenty-five years later, these statements are still timely. One can hope that the Department of Electrical and Computer Engineering at the University of Rochester will continue with a faculty that chooses to be part of interdisciplinary research. The setting is unique and the associations have proved to be productive and satisfying—and worthy of recording.

In literally the final days of the decade, precisely November 30, 1999, a lengthy patent infringement case involving the Blue Noise Mask was finally resolved. The settlement provided a substantial royalty stream that would be used to create endowed Professorships of Electrical Engineering. Consequently, the department would begin the twenty-first century with two distinguished professorial chairs, one awarded to

Professor Murat Tekalp and one to Professor Eby Friedman, thus bringing the authorized department faculty strength up to seventeen.

Faculty at the beginning of the twenty-first century:

- Professors Alexander Albicki, Mark F. Bocko, Philippe M. Fauchet (chair), Marc J. Feldman, Eby G. Friedman, Thomas Y. Hsiang, Thomas B. Jones, Charles W. Merriam, Kevin J. Parker (Dean), Roman Sobolewski, A. Murat Tekalp, Edward L. Titlebaum, and Robert C. Waag.
- Associate Professors Alan M. Kadin and Jack G. Mottley.
- Assistant Professors David Albonesi, and Diane Dalecki.
- Senior Lecturer Victor V. Derefinko.
- Professors Emerti Edwin L. Carstensen, Lloyd P. Hunter, Edwin Kinnen, and Sidney Shapiro.
- Visiting Professors David Blackstock and Leonid Tsybeskov.
- Adjunct Professors Chang Wen Chen, A. Tanju Erdem, E. Carr Everbach, Michael Kriss, Alexander Pentland and Eli Saber.

They would soon be joined by Wendi Heinzelman (Ph.D. MIT, 2000) and Martin Margala (Ph.D. Alberta, 1998) who would expand department investigations into the problems of wireless communication and low power integrated circuits.

> Our road is not built to last a thousand years, yet in a sense it is. When a road is once built, it is a strange thing how it collects traffic, how every year as it goes on, more and more people are found to walk thereon, and others are raised up to repair and perpetuate it and keep it alive.
>
> *Robert Lewis Stevenson*

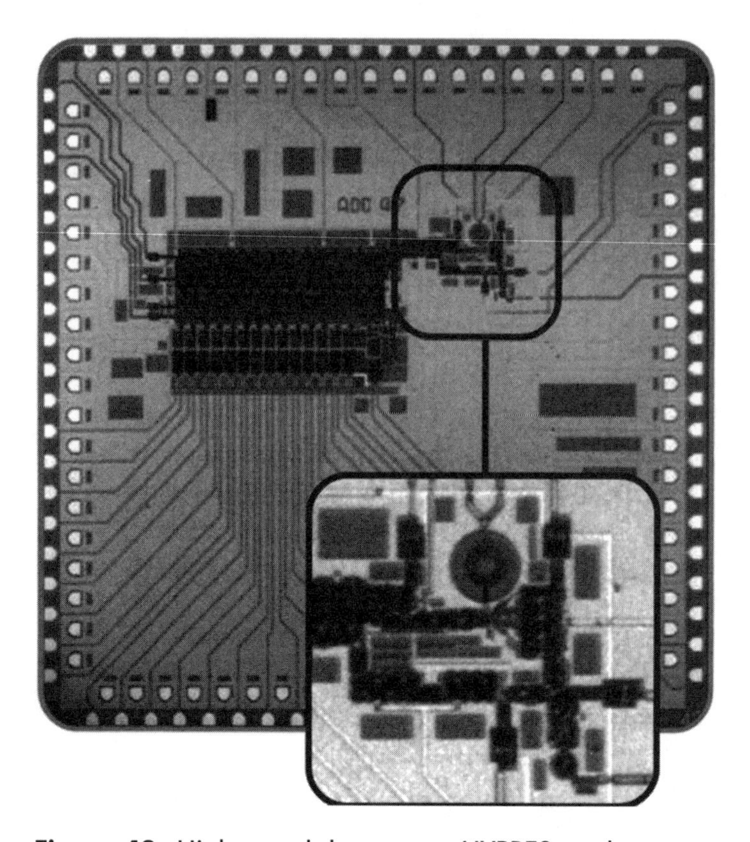

**Figure 48.** High speed, low power HYPRES analog to digital converter with a 10 Ghz RSFQ ring oscillator clock developed in Professor Bocko's laboratory, 2000.

# 7: Closing Thoughts

Every undergraduate student in electrical engineering soon learns that resonance happens when an inductor and a capacitor are connected in a simple circuit and the circuit is excited. Later, these students also learn that by carefully sensitizing a circuit to a particular frequency—that is, designing a resonant circuit for a high Q—a very small signal can excite a very large response. Suggesting an analogy between a resonant circuit and an academic department is a bit of a stretch. But for extraordinary things to happen within a department, clearly the right circumstances must be there along with a small starting event.

The preceding pages attempt to show the stimuli and the special conditions that worked together from the (second) beginning of the department. Electrical engineering instruction began at an opportune time, after World War II when electrical technology was expanding rapidly and significantly beyond the rather staid era preceding. The size of the department faculty was relatively small in a university that was also relatively small. This forced decisions by the faculty to cover less material in the classroom than would be required to reflect the broad field of electrical engineering. The economies of large classes were usually not an issue and otherwise kept at bay. Together, this allowed the faculty time for their research, and the results were especially noteworthy. The campus layout and strong departments in engineering, science, and medicine close by facilitated personal interactions. In this setting, combinations of like-minded individuals developed and started what often became interdisciplinary research programs. These cooperative efforts proved to be far more productive and long-lasting than would be expected from isolated programs by the same individuals. Added benefits occurred when new insights and applications resulted from cross-disciplinary studies.

Research projects in the department not infrequently were related to products under development in local companies competing for market share on the national and international scene. Contacts with these organizations, including the Eastman Kodak Company, the Xerox Corporation, the Gleason Corporation, R.F. Communications Division of the Harris Corporation, General Electric Medical Systems, and Bausch and Lomb, Inc., encouraged scholarly work in the community beyond the edges of the university. Informal conversations and professional meetings, brought about through shared interests, helped to keep studies in the department focused and adapted to new developments. There was also a small but significant interchange of researchers, some joining the department as new members of the faculty, others leaving the department for positions in local industry.

To a degree, the process of collecting the individuals who would join the department involved some randomness. Nevertheless, their personalities and collegial style were also some of the right components for research programs to resonate in the department, as in biomedical engineering and superconductivity. Close ties were made with colleagues elsewhere and the consequences of the research continued outside the lives of the originators, both at the university and beyond.

The freedom and commensurate time to pursue individual interests was, in a sense, the small signal that allowed the development of the research that thrived in the department. Unfortunately, time to pursue individual interests is often a minor component in the decision processes used by administrators in academia struggling to reduce costs and do more with less. This freedom and time was evident in the department, however, and the response was rewarding many times over and in many ways.

In a more subjective sense, resonance also implies vibrancy, mellowness, and fullness. Looking back over forty years, this history shows that out of the specifics, the particulars of electrical circuits and obtuse mathematical analyses, the Department of Electrical and Computer Engineering became a vibrant and exciting place for a new young faculty. Intellectual curiosity, research, instruction, and graduate student mentoring merged, and the results of this merger continue to resonate to this day.

> "Ah," sighed the melamed (teacher of Hebrew to young children in the shtetl), "if I were a Rockefeller, I'd be richer than Rockefeller."
> "How could that be?"
> "I would do a little teaching on the side."[1]

# Post Script

It is sad indeed to note here that Zhe Zheng, who received his B.S. degree major in Electrical Engineering in 1995 and an MBA in 1998, was killed in the World Trade Center disaster on September 11, 2001. Quoting from the *Rochester Review*, Winter 2002:

> Zhe Zheng could have stayed safely in the offices of the Bank of New York following the September 11 terrorist attacks. Blocks away from the World Trade Center, he was in no danger. But Zheng, a trained emergency medical technician who volunteered for the Brighton ambulance service while a student at Rochester, went to see if he could help those who were not safe.
>
> He was last seen in grainy video footage leaning over a prostrate body in the rubble of the World Trade Center. "It didn't surprise anyone who knew him," his supervisor, Peggy Farrell, told *The New York Times*. "He was a completely selfless person—he was just someone who would automatically volunteer his assistance. To me it was a truly heroic display."

**Figure 49.** Zhe Zheng.

# Appendix A: Faculty of the Department Elected Fellows to The Institute of Electrical and Electronic Engineers, as of the year 2000

Lloyd P. Hunter, 1967
For contributions to semiconductor research and publications
Charles W. Merriam, III, 1984
For contributions to optimal control theory and its applications
Edwin L. Carstensen, 1985
For contributions to understanding dielectric and ultrasound properties of biological materials
Sidney Shapiro, 1987
For contributions to superconductivity research and the discovery of the ac Josephson effect
Robert C. Waag, 1990
For contributions to biomedical ultrasound in the areas of tissue characterization, wave propagation effects and two-dimensional imaging
Thomas B. Jones, Jr., 1995
For contributions to the electromechanics of particles, and their control and manipulation using electric and magnetic fields
Kevin Parker, 1995
For contributions to the advancement of medical ultrasound, including the development of sonoelasticity imaging and ultrasound contrast agents
Philippe M. Fauchet, 2000
For contributions to nanoscale silicon optoelectronics
Eby G. Friedman, 2000
For contributions to high performance circuit design and VLSI-based synchronous systems

**Figure 50.** Aleksandar Jovancevic, left, and Zoran Mitrovski with their new doctorate diplomas, 1999.

# Appendix B: Ph.D. Degrees, 1962–1999

**1965**

Julius L. Goldstein (Professor Voelcker)
An Investigation of Monaural Phase Perception
John Crampton Heurtley (Professor Streifer)
A Theoretical Study of Optical Resonator Modes and a New Class of Special Functions
David Platnick (Professor Cohen)
Study of a Statically Switched D.C. Powered Motor
Leon L. Wheeless (Professor Cohen)
The Effects of Intensity on the Eye Movement Control System

**1966**

Lee E. Ostrander (Professor Cohen)
The Use of Inner Product Concepts in Obtaining Process Representations
Robert E. Lee (Professor Cohen)
Effects of Adaptation Level and Stimulus Amplitude on Latency to Contraction of the Human Pupil Reflex
Anthony Irving Eller (Professor Flynn)
Dynamics of a Translation Cavity in a Liquid
Robert Anthony Rubega (Professor Flynn)
The Determination of Sound Velocity Gradients in the Ocean by Acoustic Means

**1967**

Clark Nelson Kurtz (Professor Streifer)
A Theoretical Study of Guided Submillimeter Through Optical Waves in Inhomogeneous Media
Tzu-Fann Shao (Professor Hunter)
A Study of the Physical Mechanism of the Chargistor
John G. Webster (Professor Cohen)
Temporal Factors in the Pupillary Light Reflex System

**1968**

Chiou-Shiun Chen (Professor Kinnen)
Liapunov Function Generation for Autonomous Systems
W. Bromley Clarke (Professor Cohen)
Static and Dynamic Characteristics of Carotid Sinus Baroreceptors
Charles William Einolf, Jr. (Professor Carstensen)
The Low Frequency Dielectric Disperson of Micro-organisms
Roy Lee Testerman (Professor Healy)
Reflex Control of Vocalization in Cats
John Stewart Thompson (Professor Titlebaum)
Radar Signal Design Using the Ambiguity Function
Thomas J. Walter (Professor Gamo)
Instrumentation for Higher Order Statistics of Radiation and Experimental Study for Superradiant Radiation

**1969**

Lawrence L.C. Chik (Professor Cohen)
Applications of Self-Adjoint Differential Systems
Avinash Ramkrishna Karnik (Professor Cohen)
Optimal Synthesis of Distributed Parameter Systems
George Takashi Koide (Professor Carstensen)
The Dielectric Properties of Structured Water
Robert Lawrence Sanderson (Professor Streifer)
The Calculation of Laser Resonator Modes
James William Sinko (Professor Streifer)
A New Mathematical Model for Describing the Age-Size Structure of a Population of Simple Animals

Marc Steven Weiss (Professor Ruchkin)
The Effects of Changes in Conditioned Behavior on Some Properties of the Evoked Potential

**1970**

Donald B. Cruikshank, Jr. (Professor Blackstock)
An Experimental Investigation of Finite-Amplitude Oscillations in a Closed Tube at Resonance
Aristides A.G. Requicha (Professor Voelcker)
Contributions to a Zero-Based Theory of Bandlimited Signals
Ryuzo Yokoyama (Professor Kinnen)
The Inverse Problem of the Optimal Regulator

**1971**

Richard A. Altes (Professor Titlebaum)
Methods of Wideband Signal Design for Radar and Sonar Systems
Shanti P. Chakravarty (Professor Kinnen)
Frequency Domain Stability Criteria for Nonlinear Systems
Shih-Shung Chuang (Professor Gamo)
Infrared Isolator Using Yttrium Iron Garnet and Xenon Unilateral Superradiant and Radiation
Giora Gafni (Professor Kinnen)
Pulsatile Pulmonary Venous Blood Flow in Dogs
Robert Edmond Grace (Professor Gamo)
Statistical Measurement of Intensity Fluctuations in Single-Mode Gas Lasers Near the Oscillation Threshold
Clark Allen Hamilton (Professor Shapiro)
RF-Induced Effects in Josephson Tunnel Junctions
Anil Kumar Jain (Professor Cohen)
Synthesis, Identification and Control of Distributed Parameter Systems
Charles W. Newman (Professor Cohen)
An Investigation of the Human Saccadic Visual Tracking System
K. Bradley Paxton (Professor Streifer)
An Analytic Geometric Optics Study of Rays in Radially Inhomogeneous Guiding Media

## 1972

Richard Alan Johnson (Professor Titlebaum)
Energy Spectrum Analysis as a Processing Mechanism for Echolocation
Frederick William Kremkau (Professor Carstensen)
Macromolecular Interaction in the Absorption of Ultrasound in Biological Material
Kenneth Kun-E Park (Professor Kinnen)
Stability and Optimal Control of Distributed Parameter Systems
Gerrit Lubberts (Professor Shapiro)
Electron Tunneling Through Thin Semiconductor Films
Richard Andrew Rink (Professor Streifer)
Applications of a Digital Computer to Solve Analytically Special Classes of Linear and Nonlinear Differential Equations

## 1973

Mehdi N. Araghi (Professor Das)
Generation and Amplification of Ultrasonic Layer Waves
Michael L. Daley (Professor Cohen)
An Investigation of the Human Horizontal Tracking System
Joan Rose Ewing (Professor Hunter)
A Study of the Surface Properties of Silicon
Andrew Longacre, Jr. (Professor Hunter)
Josephson Frequency Conversion in Superconducting Point-Contact Junctions
Frederick R. Ruckdeschel (Professor Hunter)
Dynamic Contact Electrification Between Insulators

## 1974

Isaac Ishola Ajewole (Professor Waag)
Maximum Likelihood Receivers for Nonlinear Digital Communication Systems
Rakesh Kumar (Professor Hunter)
Collector Capacitance and High Level Injection Effects in Bipolar Transistors

Yu-Hwan Lin (Professor Kinnen)
Construction of Liapunov Functionals for Distributed Parameter Systems

**1975**

David Elmer Nelson (Professor Titlebaum)
A Statistical Scattering Model for Time-Spread Sonar Targets

**1976**

Leon A. Frizzell (Professor Carstensen)
Ultrasonic Heating of Tissues
Tesleem Iyola Raji (Professor Waag)
Phenomenological Characterization of Transmission—Reflection Processes in Diagnostic Ultrasound
Charles V. Stancampiano (Professor Shapiro)
Injection-Locking Phenomena in Josephson Oscillators

**1977**

James Lacey Cambier (Professor Wheeless)
Binucleate Cell Recognition and Prescreening System Design in Automated Gynecologic Cytopathology

**1978**

Paul Poo-Kam Lee (Professor Waag)
Noninvasive Ultrasonic Blood Flow Characterization
Robert Marc Lerner (Professor Waag)
Diffraction-Based Characterization of Biological Tissue with Ultrasound
Sankaran Narayana Murthy (Professor Titlebaum)
Sequential Least Squares Estimation of Slowly Time-Varying Parameters

**1979**

Vasant Durgadas Saini (Professor Cohen)
Use of the Pupil to Study the Scotopic and Chromatic Mechanisms of Vision in Humans

**1981**

Robert Bruce Tilove (Professor Voelcker)
Exploiting Spatial and Structural Locality in Geometric Modelling

**1983**

Maciej J. Ciesielski (Professor Kinnen)
Digraph Relaxation for Computer Aided Generation of Custom IC Layout

**1984**

James Andrews Campbell (Professor Waag)
Measurements of Ultrasonic Differential and Total Scattering Cross Sections for Tissue Characterization
Christopher Valentine Hollot (Professor Barmish)
Construction of Quadratic Lyapunov Functions for a Class of Uncertain Linear Systems
Ian Richard Petersen (Professor Barmish)
Investigation of Control Structure in the Stabilization of Uncertain Dynamical Systems

**1985**

Krzysztof Antoni Kozminski (Professor Kinnen)
Rectangular Dualization for Computer Aided Design of VLSI Circuit Floorplans

Jaroslaw R. Rossignac (Professor Requicha)
Blending and Offsetting Solid Models

**1986**

Ertugrul Berkcan (Professor Kinnen)
Performance Biased Placement for Integrated Circuit Layout Design
Donald Philip Butler (Professor Hsiang)
Nonequilibrium Behavior of the Dynamic Resistive Transition in Superconducting Thin Films
Todd A. Jackson (Professor Albicki)
Metastable Operation of NMOS Flip-Flops
Ronald N. Yeaple (Professors Carstensen and Kinnen)
Botulinum Toxin Therapy and the Oculomotor Control System

**1987**

Stephen Chi Fai Chan (Professors Voelcker and Tourassis)
MPL: A New Machining Process/Programming Language
Douglas Raymond Dykaar (Professor Hsiang)
Picosecond Switching Measurements of a Josephson Tunnel Junction
Charles Surya (Professor Hsiang)
Flicker Noise in P-Channel Metal-Oxide Semiconductor Field-Effect Transistors

**1988**

Marcelo Huibonhoa Ang, Jr. (Professor Tourassis)
Kinematic Control of Robotic Manipulators with Multiple Interchangeable Wrists
Eveline Jeannine Ayme (Professor Carstensen)
Transient Cavitation Induced by High Amplitude Diagnostic Ultrasound

Jerome Rene Bellegarda (Professor Farden)
Recursive Parameter Estimation for Time-Varying System Identification
Zeynep Çelik Butler (Professor Hsiang)
1/f Noise in Semiconductor Devices (MOSFET)

**1989**

James Anthony Beausang (Professor Albicki)
Test Control for Self-Testable VLSI Chips
Justin Christian Goding, Jr. (Professor Farden)
Matrix Product Inequalities for Convergence Analysis
Badri Lokanathan (Professor Kinnen)
Floor Plan Design: Connectivity-Driven Transformations from System Representation to Rectangular Dual
Mario Merri (Professor Titlebaum)
Static and Dynamic Analyses of Ventricular Repolarization Duration
Ruth Douglas Miller (Professor Jones)
Frequency-Dependent Orientation of Lossy Dielectric Ellipsoids in AC Electric Fields
Louis F. Pizziconi (Professor Waag)
Ultrasonic Propagation in a Weakly Inhomogeneous Medium
Robert Hammond Sperry (Professor Parker)
Segmentation of Speckle Images Based on Level Crossing Statistics
John F. Whitaker (Professor Hsiang)
Ultrafast Electrical Signals: Transmission on Broadband Guiding Structures and Transport in the Resonant Tunneling Diode

**1990**

Mondher Ben-Ayed (Professor Merriam)
Dynamic Routing for Regular Direct Computer Networks
Arif Tanju Erdem (Professor Tekalp)
Modeling Higher-order Spectra with Applications in Signal and Image Processing
Sung-Rung Huang (Professor Parker)
Principles of Sonoelasticity Imaging and its Applications in Hard Tumor Detection

Svetislav Vojislav Maric (Professor Titlebaum)
New Frequency Hop Coding Methods with Applications
Jan Hendrik Vandenbrande (Professor Requicha)
Automatic Recognition of Machinable Features in Solid Models
Xing Zhou (Professor Hsiang)
Ensemble Monte Carlo Modeling of High-field Transport and Ultrafast Phenomena in Compound Semiconductors

**1991**

Rajiv Arora (Professor Albicki)
Prioritized Access in Carrier-sense Multiple Access Local Area Networks
Dimitrios Emiris (Professor Tourassis)
Kinematic Analysis, Evaluation and Control of Dual-elbow Robotic Manipulators
Evagelos Katsadas (Professor Kinnen)
Multi-layer Routing for General Cell Layouts Utilizing Over-the-cell Areas
Theophano Mitsa (Professor Parker)
Digital Halftoning Using a Blue Noise Mask
Robert Francis Molyneaux (Professor Albicki)
Built-in Test Design for the Efficient Testing of VLSI Circuits
Kit-Ming Wendy Tang (Professor Arden)
Dense-Symmetric Interconnection Networks
Theresa Ann Tuthill (Professor Parker)
Dominant Mechanisms of Ultrasound Tissue Interactions in Liver

**1992**

Albert Tatum Davis (Professor Bowman)
Implicit Mixed-mode Simulation of VLSI Circuits
Zoran Ilija Kostic (Professor Titlebaum)
Spread Spectrum Coding, Performance Evaluation and Dispersive Channel Estimation for Several Classes of Code-Division Multiple-Access Communication Systems
Sanjay Kumar Mehta (Professor Titlebaum)
Signal Design Issues for the Wigner Distribution Function and a New Twin Processor for the Measurement of Target and/or Channel Structures

Mehmet Kemal Ozkan (Professors Tekalp and Sezan)
Multiframe Filtering Techniques for Video Signal Processing
Gordana Miroslav Pavlovic (Professor Tekalp)
Identification and Restoration of Images Based on Overall Modeling of the Imaging Process
Thomas Nathaniel Tombs (Professor Jones)
The Effect from Absorbed Water on the Polarization of Individual Glass Particles Suspended in Silicone Oil (a Model Electrorheological Fluid)

**1993**

P. Ted Christopher (Professor Parker)
Modeling Acoustic Field Propagation for Medical Devices
Diane Dalecki (Professor Carstensen)
Mechanisms of Interaction of Ultrasound and Lithotripter Fields with Cardiac and Neural Tissues
Michael Aaron Fisher (Professor Bocko)
Development of a Radio Frequency, Superconducting, Electromechanical Transducer
Bin Liu (Professor Wengler)
2-D Quasi-optical Josephson Junction Arrays for THz Oscillators
Aleksandar Pance (Professor Wengler)
Two-dimensional Josephson Arrays for Submillimeter Coherent Sources
Zhao-Nan Zhang (Professor Bocko)
A Method to Circumvent Quantum Noise Limits in Superconducting Tunnel Junction Mixers

**1994**

Sotiris Alexandrou (Professor Hsiang)
The Bent Coplanar Waveguide at Sub-terahertz Frequencies
Fabio Francesco Badilini (Professor Titlebaum)
Time and Frequency Analysis of ST Segment Displacement Signals in Ambulatory ECG Recordings

Paul Henry Ballentine (Professor Kadin)
High Temperature Superconducting Thin Films: Sputter Deposition and Fast Optical Switching
Noshir Behli Dubash (Professor Wengler)
Photon Induced Noise in Superconducting Tunnel Junction Detectors
Brenda L. Luderman (Professor Kinnen)
A Critical Analysis of Modeling and Simulating Submicron Semiconductor Devices Using Classical Transport Theory and the Finite Element Method
Gordana Pance (Professor Wengler)
Broadband Integrated Tuning Circuits for Quasioptical SIS Mixers: Design and Performance
Jonathan Karl Riek (Professor Tekalp)
Modeling, Detection and Suppression of Motion Artifacts in Magnetic Resonance Imaging
Bob Srbislav Stanojevich (Professor Bowman)
Automated Cell Synthesis of Analog Integrated Circuit Layout ANASYN
Kendall Alvin Stephenson (Professor Bocko)
The Vacuum Tunneling Probe as a Nonreciprocal Quantum Limited Transducer
Bradford Clark Tousley (Professor Fauchet)
Ultrafast Carrier Dynamics of Bulk InGaAs, Low Temperature Grown InGaAs, and MQW InGaAs
Xiandong Xie (Professor Albicki)
Constructive Techniques for Delay Fault Testing

**1995**

Brian S. Cherkauer (Professor Friedman)
CMOS-based Architectural and Circuit Design Techniques for Application to High Speed, Low Power Multiplication
Lan Gao (Professor Parker)
Sonoelasticity: Theory and Experiment Development
Razak Hossain (Professor Albicki)
Low Power CMOS Circuits Through Reduced Switching Activity
Qing Ke (Professor Feldman)
Superconducting Single Flux Quantum Circuits Using the Residue Number System

Derek Scott Mallory (Professor Kadin)
Active Microwave Bandpass Filters Using High-temperature Superconductors
Stephen Swanston Martinet (Professor Bocko)
An Investigation of a New Superconducting Logic Family: Design and Experimental Low-Speed Testing of its Circuits
Kumar Neelakantan (Professor Mottley)
Modeling and Measurement of Ultrasonic Backscatter and Attenuation from Myocardial Tissue

**1996**

Sheikh Kaisar Alam (Professor Parker)
The Butterfly Search Blood Velocity Estimation Technique for Doppler Ultrasound Flow Imaging
Edward Andrew Ashton (Professor Parker)
Segmentation and Feature Extraction Techniques, with Application to Biomedical Images
Michael Ming Hsin Chang (Professors Tekalp and Sezan)
Bayesian Analysis and Segmentation of Multichannel Image Sequences
Hei Tao Fung (Professor Parker)
Efficient Segmentation and Compression of Scanned Document Images
Deepnarayan Gupta (Professor Kadin)
Optically Triggered Superconducting Opening Switches
Jiebo Luo (Professor Chen)
Low Bit Rate Wavelet Based Image and Video Compression with Adaptive Quantization, Coding and Postprocessing
Jose Luis Pontes Correia Neves (Professor Friedman)
Synthesis of Clock Distribution Networks for High Performance VLSI/ ULSI-based Synchronous Digital Systems
Andrew John Patti (Professor Tekalp)
Digital Video Filtering for Standards Conversion and Resolution Enhancement
Cheng Peng (Professor Fauchet)
Studies of Light Emitting Porous Silicon
Eli Said Saber (Professor Tekalp)
Automatic Image Annotation and Query-by-example Using Color, Shape and Texture Information

Christopher Wayne Summers (Professor Mottley)
    On the Reconstruction of Subresolvable Point Scatterers in Pulse-Echo Imaging: A Theoretical Study
Wei Xiong (Professor Sobolewski)
    Fabrication and Optoelectronic Properties of Y-Ba-Cu-O Thin Films with Different Oxygen Contents
Meng Yao (Professor Parker)
    Blue Noise Halftoning

**1997**

Yucel Altunbasak (Professor Tekalp)
    Object-scalable, Content-based Video Representation and Motion Tracking for Visual Communications and Multimedia
Darren Keith Brock (Professor Bocko)
    Design Methodologies for Mixed Signal Multi-gigahertz RSFQ Circuits
Mark David Hahm (Professors Titlebaum and Friedman)
    Receiver Issues Related to Spread Spectrum Communications
Linda Elizabeth Marchese (Professor Bocko)
    Multimode Electromechanical Transducers and Optimum Filters for Resonant Mass Gravitational Antennas
Chak L. Tan (Professor Jones)
    Lattice Granular Electromechanics
Chia-Chi Wang (Professor Hsiang)
    Ultrafast Testing of Electronic/Optoelectronic Devices
Xiaohui Wang (Professors Ning and Parker)
    Volume Tomographic Angiography

**1998**

Laura Burattini (Professor Titlebaum)
    Electrocardiographic T-wave Alternans Detection and Significance
Siddhartha Prakash Dutta Gupta (Professor Fauchet)
    Porous Silicon Based Optoelectronics: Characterization, Process Integration, and Device Fabrication
Quentin Paul Herr (Professor Feldman)
    Bit Errors and Yield Optimization in Superconducting Digital Single-Flux-Quantum Electronics

Douglas W. Jacobs-Perkins (Professor Hsiang)
Design, Analysis and Implementation of an Ultrafast Electro-Optic Electric-Field Imaging System
Zoran Mitrovski (Professor Titlebaum)
Multipath Channel Deconvolution and Variable Bit-Rate Methods for Spread Spectrum Communication Systems
José Ernesto Nuñez-Regueiro (Professor Kadin)
Magnetic Field Effect Devices Using Perovskite Thin Films
Daniel Brian Phillips (Professor Parker)
Simulation of Ultrasonic Scattering from a Fractal Model of the Liver
Candemir Toklu (Professors Tekalp and Erdem)
Object-based Digital Video Processing Using 2-D Meshes
Kamil Burak Ucer (Professor Fauchet)
Ultrafast Carrier Dynamics in Thin Porous Silicon Films
Xiaozhen Weng (Professor Fauchet)
Ultrafast Spectroscopic Studies of Conjugated Polymers
Jian Wu (Professors Fauchet and Kadin)
Development of Infrared Detectors for Space Astronomy
Jianbin Wu (Professor Albicki)
Adiabatic Charging, Reversible Logic and Low Power Circuits
Fai Yeung (Professors Levinson and Parker)
Motion Estimation and Analysis of Ultrasound Image Sequences
Qing Yu (Professor Parker)
Quality Issues in Blue Noise Halftoning

**1999**

Victor Seth Adler (Professor Friedman)
Repeater Insertion for Driving Resistive Interconnect in CMOS VLSI Circuits
Marc Daniel Currie (Professor Hsiang)
Ultrafast Electro-Optic Testing of Superconducting Electronics
Kenton Andrew Green (Professor Sobolewski)
Characterization of Time and Frequency Using Optoelectronic Microwave Silicon Switches
Aleksandar Velimir Jovancevic (Professor Titlebaum)
Analysis of Frequency Hopping CDMA Systems with Emphasis in New Coding Families and Improved Receiving Techniques

Ivan Stefanov Kourtev (Professor Friedman)
Enhanced Algorithms for Non-Zero Clock Skew Scheduling
Cesar Augusto Mancini (Professor Bocko)
Short-term Frequency Stability of Clock Sources for Digital Single-flux-quantum Electronics
Lisa Ann Osadciw (Professor Titlebaum)
Synchronization of Spread Spectrum Communication Systems Using Signal Design
Jeffrey A. Small (Professor Parker)
Conjugate Channel Transmission of Multitone Images
José Gerardo Tamez-Peña (Professor Parker)
Four Dimensional Reconstruction and Visualization of Complex Musculoskeletal Structures
Nada Vukovic (Professor Feldman)
Residue Number System Arithmetic Implementation in Superconducting Single-flux-quantum Digital Technology

**Figure 51.** Professor Parker, left, with Master's recipient, Vladimir Misic, 1999.

# Appendix C: M.S. Graduates

**1953**

Robert Q. Pollard

**1958**

Stanley J. Dudek
Lorand W. Magyar-Wilczek
Ronald G. Matteson
Clor William Merle

**1959**

Joseph A. Huie
Peter H. Zachmann

**1960**

Miles Davis
James L. Douglas
Clifford P. Oestreich

**1961**

Arthur W. Alphenaar
George Arthur Brown
Robert C. Curry

Thomas E. Hattersley, Jr.
Vladimir Kushel
David Platnick
Thomas Proctor
Raymond J. Rogers
Menashe Simhi

**1962**

Robert H. Aronstein
W. Bromley Clarke
Michael K. Karsky
Robert E. Lee
Theodore B. Mehlig
Arthur Raymond Phipps
Frederick G. Reinagel
Robert A. Rubega

**1963**

Bertrand E. Berson
Edward S.I. Chiang
Anthony Eller
J. Warren Gratian
Robert Frederick Osborne
Lee E. Ostrander
Kodati Subba Rao
James R. Verwey
Leon L. Wheeless, Jr.
Walter W. White

**1964**

Chiou-Shiun Chen
William C. Evans
Alfred Frederick Gaspar
Chao-Peng Hsieh

Alexander E. Martens
James C. Runyon
Robert L. Sanderson
Kenneth G. Shepherd
James H. Taylor
Roy L. Testerman
Kenneth D. Urfer
Thomas J. Walter
Chung-Chian Wang
Marc S. Weiss
Douglas H. Williams
Joseph H. Worth

**1965**

James K. Bannon
Charles W. Einoff, Jr.
Martin Bennett Gray
Robert Raymond Heath
Robert A. Houde
Michael H. Kalmanash
John E. Keenan
Karl H. Kostusiak
David Alan Naus
Koho Ozone
Charles W. Newman
Robert Buch Renbeck
Noel L. Reyner
John S. Thompson
George Lemar Thompson, Jr.
John G. Webster

**1966**

Richard A. Altes
Donald B. Cruikshank, Jr.
Leo Thomas Gogowski
Stanley Jens Haavik

Lawrence Arnold Klein
William Robert Parry
K. Bradley Paxton

**1967**

Shanti P. Chakravarty
Lawrence L. Chik
Shih-Shung Chuang
John Lyman Connin
Joan R. Ewing
Ernest J. Laszio
Gerrit Lubberts
J. Terrence Montonye
Karnam Rameswar Rao
Aristides A.G. Requicha
Richard A. Rink
Murray Shostak
Gerald Leslie Sperber
Yehoshua Y. Zeevi

**1968**

Chiou-Shiun Chen
Guy Peter Clark
Richard Alan Garlick
Clark A. Hamilton
Gilbert W. Herzog
Richard A. Johnson
Peter Freemman Lemkin
Rafael Llavina, Jr.
Andrew Longacre, Jr.
James Wallace Luening
Arnold Mark Michelson
Michael F. Salata, Jr.
Walter E. Shepherd
Alan Joseph Stankus

Thomas Laszio Szabo
John S. Thompson
William Stanford Walsh
Joseph P. Zigadio

**1969**

J.S. Agarwala
Prathima Arakere
Nausheer Jal Avari
Steven Charles Bergman
Bruce M. Cleveland
John Karl Elberfeld
Orin Queal Flint, Jr.
Anshoo Sudhir Gupta
Anil K. Jain
Edward H. Kleiner
Frederick W. Kremkau
Michael Gerald O'Donnell
Kenneth K. Park
Alton F. Riethmeier
Richard Alan Rosner
James C. Runyon
Peter Alan Saltz
Rajesh Saxena
Jun-Been Shao
Joseph Stasko
William John Taylor
Gerald Francis Woolever
Martin Ross Zalonis

**1970**

Manuel Enrique Arellano-Ramirez
Yi-Hua Edward Chang
Jose Luis Delgado
David E. Kellock
Dev Karan Khanna

Jacques Paccard
Richard Olney Rhodes
Madhu Sudhana Rao Sreepada
Andrew J. Steckl
Kai-Ping Yiu

**1971**

Bela Bessenyei
Leon A. Frizzell
Dilip Pandurang Joshi
Frederick R. Ruckdeschel
Karl Egon Schertler
Daljeet Singh
Charles V. Stancampiano
Ashok Sud

**1972**

Russell Carl Bogardus
James L. Cambier
Peter G. Formaniak
David Charles Hogan
Venkatesh H. Kamath
Yu-Jih Liu
Thomas Benjamin Michaels
Samkaran N. Murthy
Joel Bruce Myklebust
Carol Ann Niznik
Willard J. Rhoads
Ronald Joseph Sudol
Hsing Chen Tuan

**1973**

Bruce Robert Benwood
Maghar Singh Chana

Frederick Allyn Eames
Eric Mitchell Geller
William Joseph Haller
Roupen H. Maronian
Peter T. Palamar
David Russell Smith
John Somerville Snyder, Jr.

**1974**

Dilip Madhav Apte
Gary Kean Budd
Thomas Anthony Deletto
Daniel Herbert Emrick
Sun-Yuan Kung
Regis Gary Lagler
Paul Poo-Kam Lee
Pok Ming Leung
Vijay Luthra
Takashi Shigihara
Shamal L. Suthar
Victor Tawil

**1975**

Kah-Fae Chan
Azim Alibhai Velji Dosani
Gideon Gimian
Paul Vincent Indaco
Stjepan A. Jarnjak
Francis Kwokyiu Ma
Vasant D. Saini
Myles Arthur Shaftel
Himanshu Bhogilal Shah
Jose Manuel Sierra
Raymond Chi Sin Tzau
David Edward Zaucha

**1976**

Eugene Stephen Evanitsky
Ter-Tsu Huang
Philip Francis Hurt
Etsuo Iwakami

**1977**

Bruce Raymond Hutchison
Devendra Kaira
Marana Anne Kern
Joseph Lanzillotta
Jeffrey Stephen Pidel
Kenneth Arthur Shulman

**1978**

Farrokh Abrishamkar
Steven Allen Becroft
N. Kirk Birrell
Dale G. Lakomy
John Lucian McClaine
Bruce Alan Peterson
Robert James Pizzutiello, Jr.
Neil Eric Rich
James Paul Shipkowski
Mark Wesley Smith
Robert B. Tilove

**1979**

David Mark Balzer
James Andrews Campbell
Yuk Yuen Chen
Adam Joshua Efron
W. Burns Fisher
Justin C. Goding, Jr.

Daniel Steven Kinney
Ashok Jaikishen Rajpal
Khalid Sayood
Thomas Steven Soulos
Lyle M. Tague

**1980**

William Daniel Auman
David Alan Butler
Dipankar Ghosh
Deborah Ronnie Horowitz
James Clinton Maher
Ian Petersen
Paul A. Rulli
William M. Speth, Jr.
George W. Swartout

**1981**

Donald Philip Butler
Patrick S. Chan
Douglas Raymond Dykaar
Christopher V. Hollot
Takuya Imaide
Ujjaldip Singh Kohli
Douglas Mitchell Schwarz
Robert Sperry
John Robert Thompson

**1982**

Mahmoud Abdallah
Antoun I. Ateya
Thomas J. Crawford
Todd A. Jackson
Bonnie Ann Maye

Amitabha Mukhopadhyay
Gregory C. Sosinski
Kehui Wei

**1983**

Peter John Delfyett, Jr.
Alberto Ricardo Galimidi
Gregory James Martin
Heiko Rommelmann
Jaroslaw Roman Rossignac
Jacques Stroweis
Charles Surya
John Firman Whitaker

**1984**

Eveline Ayme
Robert Bartlett
James Anthony Beausang
Jerome R. Bellegarda
Stephen Chi Fai Chan
Paul Edward Christopher
John William Lockard
Sharon Ann Mathiason
Robert William McClellan
William David Notovitz
Venkatanarayana Paruchuri
Michael Rosa
Paul James Rose
Vwani P. Roychowdhury
Ali Ugur Sungurtekin
Jan Hendrik VandenBrande

**1985**

Rajiv Arora
Diane Dalecki

Ruth Douglas
Jane Kira Eisen
Charles Eleiott
Scott M. Hayman
Michael Allen Inchalik
Varsha Kapadia
Andrzej Krasniewski
Martin Parker
Donald Ernest White
Normal W. Zeck

**1986**

Stephen John Archer
Mondher Ben-Ayed
David Frederic Carlson
James E. Fowler
Daniel Alan Gray
Matt John Harris
John P. Kraybill
John Morton Lewin
Badri Lokanathan
William A. Orfitelli
Mark Robert Scheda
Fernando Torre-Sanchez
Theresa Ann Tuthill

**1987**

Marcello H. Ang, Jr.
Paul W. Ballentine
Larry William Beattie
Thomas Richard Carducci
Joe Yul Cho
James Michael Chwalek
David Merlin Fullmer
Richard Allan Gammons
Jerzy Kalinowski
Audrey Mei-Ling Khoo

Mark Edmond Lyons
Jay D. Marchetti
Steffen W. Parratt
Paul Jerome Regensburger
Julian Sinai
Kit-Ming Tang
Xing Zhou

**1988**

Bruce Allen Alexander
James Ronald Anuskiewicz
Wei Cao
Daniel Wing Hong Chan
Ji-Yong David Chung
James Philip Coene
Barbara F. Demczar
Dimitrios Emiris
Arif Tanju Erdem
David Kirtland Fork
Michael Yuri Frankel
Shantanu Gupta
James William Holmes
Sung-Rung Huang
Ann H. Infortuna
Evagelos Katsadas
Manoj Khare
Zoran I. Kostic
K.N. Kumer
Eric John Leinberg
Charles Alan Linn
Svetislav V. Maric
Christopher Ian Marshall
Jon David Masyga
Sanjay Kumar Mehta
Mario Merri
Theophano Mitsa
Kumar Neelakantan
Gordan M. Pavlovic

Thomas L. Pratt
Clarence Raymond Reilly
Kendall A. Stephenson
Tanya Lynel Rugo
Angie Mei-Ching Wan

**1989**

Michela Alberti-Merri
Sandeep Bhatia
Christopher Jon Bucci
Noshir Behli Dubash
Michael A. Fisher
Christian G.M. Heiter
Brenda Lee Luderman
Aleksandar Pance
John George Pilgrim
Adam Dale Sherer
Thomas Nathaniel Tombs
Joseph Ward

**1990**

Sotiris Alexandrou
Cynthia Jean Baron
Derrick Shawn Campbell
Fritz Francis Ebner
Timothy M. Enders
Hiroyuki Fujita
Michael A. Inchalik
Derek Scott Mallory
Stephen Swanston Martinet
Hiroyuki Nakano
Minh Nhat Nguyen
Mehmet Kemal Ozkan
Angel Palacios
Gordana Pance
Timothy John Reardon
Jonathan Karl Riek

Martin Kenneth Slawson
Yoichi Sumino
C. Wayne Summers
Brad Clark Tousley
S. Lance VanNostrand
Walter F. Wafler
Chuan-Hui Wu

**1991**

Sheikh Kaiser Alam
Robert Alan Appleby
Fabio Francesco Badilini
Bishwa Vijaya Basnet
Nikhil Celly
Ming-Hsin Michael Chang
Brian S. Cherkauer
Jose Luis P. Correia-Neves
Tracy C. Denk
David G. Faller
Maria Helguera
James Francis Herrmann
Kurt Thomas Knodt
Krista Linn Kortkamp
John Arthur McNeill
Christopher Matthew Miceli
Andrew J. Patti
Chak L. Tan
Xiaoou Tang
Leszek Dariusz Wronski
Meng Yao
Thomas Baylis Zell
James M. Ziobro

**1992**

Brian D. Brown
Allan Chi Wan Cheung

Hei Tao Fung
Deepnarayan Gupta
Ellen Veronica Harp
Razak Hossain
Rajarao Jammy
Dikran Kassabian
Zaim Khalid
Mee-Mee Lai
Sharon Anne Lorenzo
Debbie Chaofang Lun
Bruce H. Pillman
Eli F. Saber
Ronald Christopher Scinta
David I. Seah
Rita Sherman
Andrzej Sobski
Makoto Takahashi
Jian Wu
Xiandong Xie
Wei Xiong

**1993**

Victor Adler
Yucel Altunbasak
Edward Ashton
Timothy John Case
John Thomas Compton
Marc Daniel Currie
Andrew Denysenko
Hema Mohanial Dhanesha
Mark David Hahm
Andrew King Halberstadt
Douglas Jacobs-Perkins
Rajarao Jammy
Ryoichi Kanda
Qing Ke
Cesar Mancini
Linda Elizabeth Marchese

Stephen Joseph Muscato
Cheng Peng
Kurt M. Sanger
Anthony Ward Schrock
Ronald Wayne Silkman
Tolga Soyata
Bob Srbislav Stanojevich
Nada Vukovic
Rachel Xiaoyang Wang
Xiaozhen Weng
Menghui Zheng

**1994**

Christopher David Agnew
Francoise Aurtenechea
Patrick J. Borrelli
Richard C. Cliver
Yu Fang
Lan Gao
Kenton Andrew Green
Peter F. Hallemeier
Jeffrey S. Lillie
Jiebo Luo
Zoran Mitrovski
Jose Ernesto Nunez-Rugueiro
Daniel Brian Phillips
Wendy Marie Scinta
Luis Alejandro Solis
Candemir Toklu
Jeffrey Alan Weisberg

**1995**

Kenneth S. Bhella
Darren Keith Brock
Laura Burattini

Somchart Chokchaitam
Pekin Erhan Eren
Kenneth C. Henderson
Richard Kirkcaldy Hynds
Non Ingkutanon
Aleksandar Velimir Jovancevic
Voravit Kosalathip
Ivan Stefanov Kourtev
Tamara G. Krol
Robert Glen Rhode
Dmitri Simonian
Michael A. Sperber
Charles Matthew Surowiec
Whitfield Robinson Thompson
Pubudu Chaminda Wariyapola
Carlo Anthony Williams
Di Wu
Jianbin Wu
Fai Yeung
Daofa Zhang

**1996**

Gennady L. Akselrod
Ameet S. Bhattacharya
Ittibhoom Boonpikum
Phakphoom Boonyanant
Chin Hong Cheah
Konstantinos M. Dobrolis
Mark A. Gwaltney
Timothy Charles Heywood
Weiwen Lai
Hou-Sheng Lin
Alan G. MacRobbie
Basab Mukherjee
Wray E. Paul
Jeffery Mikko Russell
Ahmed Shahid
Jingqing Song

Jose Gerardo Tamez-Pena
Rong Yao
Qing Yu

**1997**

Roman Adam
Farshid Attarian
Taqi M. Baig
Ahmet Mufit Ferman
Haotao Li
Khurram Zaka Malik
Johanna May Mitchell
Brian Christopher Porter
Steven James Pratt
Vinai Roopchansingh
Lalit Saran Sarna
Jason Stephen Smith
Steven P. Thomsen
Andy Tin-Ho Wu
Wei Zhu

**1998**

Paul Samuel Bonino
Selena Chan
Shiloh Leigh Dockstader
Yue Fu
Patrick Anthony Furchill
Yehea Ismail Ismail
Feng Lin
Xun Liu
Stephen Andrew McAleavey
Nenad Nenadic
Steve Glenville Phillip
Radu Mircea Secareanu
Zhaohui Sun
Tianwen Tang

Lawrence Steven Taylor
Muge Wang
Minghui Xia
Ning Zhuang
Robert Michael Zucker

**1999**

Lei Chen
Agnim I. Cole
Sidney Christopher Laurenceau
Michelle Shi-Ying Lo
Luis Lorenzo
Elise Marie Raffan Michaels
Vladimir Misic
Sait Kubilay Pakin
Wayne Charles Pilkington
Yanhai Song
Christopher Carl Striemer
Xiangyang Tang
Bingxiong Xu
Songtao Xu
Yaowu Xu
Xingxiang Zhou
Qingyuan Zhu

**Figure 52.** B.S. class of 1997. *Top row left to right*: Michael Bristol, Chi-Kin Tang, Agnim Cole, Shiloh Dockstader, Ren-yu Zhang, Daniel Stephens, Ali Tahir, Sumit Mohan, Kerry Denvir. *Middle row left to right*: Terrance Jones, Edwin Sulaiman, Kin Ng, Asif Shah, Zhi-Jian Shen, Philip Schremmer, Daniel Petrovich, Justin Vlietstra, Kaushik Mittra. *Bottom row left to right*: Benjamin Munger, Abbas Tahir, Nabeel Sami, Kerstin Babbit, Matthew Reh, Akshat Thanawala, Anthony Boccio, Tanzeem Choudhury.

# Appendix D: B.S. Graduates

## 1949

Zygmund J. Bara
Clement O. Bossert
Ralph J. Brown
Philip J. Buchiere
Donald P. Dise
Thomas E. Doughty
Robert Joseph Hoefer
Robert Phelps Kennedy, Jr.
Ronald A. Miller
Harry R. Nickles
Joseph Phillips
Elliott I. Pollock
Walter J. Randolph
Cecil E. Scott
Bernard J. Schnacky
Robert Q. Pollard
Grosvenor S. Wich

## 1950

John V. Adkin
Wilbur E. Ault
Irwin Shaffer Booth
Edward S. Brown
Ingvar E. Eliasson
Albert Carl Giesselmann
Lewis Miner Goodrich

William G. Graeper
Robert Clayton Isemann
Richard F. Kaiser
Nicholas Lazar
William G. Nyhof
Arthur R. Pincipe
Erick Noak Swenson
Robert B. Taylor
Marvin Trott

**1951**

John Frank

**1961**

Elwyn G. Allyn
Lloyd Reginald Campbell, Jr.
Francis J. Caravaglio
Donald G. Simcox
Harold F. Staudenmayer

**1962**

Lee F. Backus
Charles K. Bowman
M. Dale Clark
Thomas G. Coleman
Gerald Lewis Freed
Theodore H. Morse
Kenneth G. Shepherd

**1963**

Michael F. Armstrong
James Joseph Ashton

Alan Bernstein, Jr.
Gerald E. Claflin
William J. McKechney
Nicholas A. Milley
Armando Scacchetti
James E. Summers
James A. Taylor

**1964**

William John Cannon
Richard Murray Davis
Jeffrey R. Duerr
Ernest H. Forman
Leo T. Glogowski
Daniel J. Healey
Stephen Glen Kamak
Lance T. Klinger
Paul Ulrich Lind
John F. Milne
Robert C. Phear
Henry I. Simon
Robert F. Steen
John F. Tiede
Harold R. Van Voorhis
David R. Wesley
Jerald Zandman

**1965**

Richard Paul Anstee
Alan Bernstein, Jr.
Joseph J. Bodmann, Jr.
Roger W. Ehrich
Christopher C.H. Graber
Dana Brand Hopkins, Jr.
Gregory J. Maier
Alfred John Maley

Donald L. Miller, Jr.
James C. Minor
Thomas Wilde Morris
Timothy J. Rahman
Rainer H. Sahmel
Timothy L. Skola
Donald A. Soderman
Paul D. Towner, III
Harry L. Trietley, Jr.

**1966**

John William Cross, Jr.
Johnson Ottawa Curtis
Ronald Frank Dans
Thomas D. Jones, Jr.
William Carl Kicherer
Ronald Wayne King
David Bruce Newton
Curtis Allen Risley
Edgar Henry Watts, Jr.
Angus Williams Westkirk

**1967**

Richard William Bacchetta
Frederic Roy Carlson, Jr.
Lawrence E. Choice
James Edward Day
David Thompson Deihl
James George King
Edwin Phillip Mueller
R. Alan Payne
Richard Herman Poppenberg
John Edward Sullivan, Jr.
Robert Steven Towsner
Donald Josiah Weikel, Jr.

**1968**

Charles Roger Alte
Manuel Enrique Arellano-Ramirez
Eric C. Anderson
Paul John Brusil
William Jerry Burstein
Bruce A. Davy
John Feldman
Ronald Alvin Johnson
Robert Hawthorne Jones
Donald George Peck
Jeffrie Warren Saunders
Wadie N. Sirgany
James Paul Venable
Donald Stuart Wyner

**1969**

Joseph J. Argento
John Hallowell Dunlap
Philip Allan Graham
E. Eugene Hartquist
Robert Lee Hernandez
William Joseph Hewitt
Frederick Alan Hussong
Arthur Scott Kegelman
Robert Lawrence Kettig
Theodore Raymond Lapp
Alan Perry Laskin
Carol A. Niznik
Robert G. Phillips
Thomas Joseph Policano
Gordon Eugene Presher, Jr.
Alfred Robert Rudolph
Howard Vipler
Robert Hall Wallis
Stephen Andrew Zahorian

**1970**

William Dunning Beebe
Peter B. Bishop
Charles Rodgers Cowan
Keith Thomas Knox
Albert Emil Manfredi
Jeffrey H. Portnoy
Frank Peter Rakoski
Angelo Frank Rella
Edward A. Riess
Ronald Elwood Rigby
John Edward Ryan
Steven David Selwyn
Richard Jennings Soderman
Spencer L. SooHoo
Paul Vincent Trainor

**1971**

Christopher Marier Bancrof
David R. Cidale
Derek A. Coulton
Jerome Thomas Dijak
Leonard Friedman
Henry Roland Harper
David F. Hayes
Arthur David Moin
Edward Riess
Kip Anthony Souza
William James Standera
Cengiz Tanverdi
Richard Frederick Troise
Paul Emile Viau

**1972**

Robert Frank Anderson
Eleazar E. James Antonini

Jack Frederick Bailey, Jr.
William Leo Bakonis
Thomas William Barry
Richard Michael Basehore
Kevin Richard Bell
William A. Chapman
Alan B. Feinberg
Wendell Burns Fisher, Jr.
Daniel Bruce Fenton
Steven Paul Geiger
Stephen Mark Hager
Victor J. Hirsch
Viet-Dzung Hoang
Bruce Godley Littlefield
Julius Orban
Bhaskarrao V. Pant
Alan Frank Raphael
David Bruce Rummer
Thomas Herman Szolyga

**1973**

Mark Alan Cohen
Thomas C. Corner
Robert Francis Flood, Jr.
Steven Galen Gilbert
James Allen Goins, Jr.
Elliott S. Greene
Dann Alan Gustavson
Peter Rodney Hayes
Alan Beckwith Hayter
Thomas Ling-Wen Kung
Gregory John Meteyer
Gregory Johnson Parks
Kenneth Neil Rosenblum
David Edward Shields
Howard J. Vogel
David Leo Weimer

**1974**

Aram Agajanian
Sheila Z. Balke
Thomas R. Balke
James C. Bennett
Douglas C. Brainard
John R. Carlson
Walter W. Delesky
John P. Delisio
William J. Dennis, Jr.
Vincent Louis Eckel
Assis Miranda Flores
Henry Hengwah Hon
Kenneth Y Huang
Willard C. Johnson
Michael S. Jones
Warren R. Jones
Gary D. Meisel
Stephen C. Pohlig
William Matthews Putney
Mary A. Smith
R. Mark Spencer
Richard J. Spiegel
Allan Strongwater
Robert B. Tilove
Raymond C. Tzau
David B. White
Barry N. Yarkoni
Peter P. Zadarlik
Andrew J. Zaremba
Cheila Zeppos

**1975**

Wing Kai Cheng
Charles Henry Clark, Jr.
Paul Bruce Firth
Dogan Gunes

Peter Halfdan Helmers
Kwok-Woon Lai
Randall Sek-Tim Lau
Michael A. McCourt
James Michael Quinn
David Alan Russo
Farhad Sadighi
Kenneth C. Stevens
Raphael Kwok-Kiu Tam
Thuy Thi Tran
Huynh Cao Trinh
Torney Mark Vanacker
Donald Leigh Watkins
Bruce Allen Watson

**1976**

Donald Anthony
David Balzer
Robin D. Becker
John Frederick Bender
John Wells Betz
R. Edward Dakin, II
Kathleen Marie Eccles
Norman Lester Eckhardt
Steven Alan Elkind
James Steward Fasoli, Jr.
Bruce R. Hutchison
Michael P. Kehoe
David Mathias Lubanko
John Douglas Marshall
Jeffrey Metzger
Eugene Cheung-Chun Ngai
Keith Andrew Sayuk
Lawrence John Stueber, Jr.
Francis Joseph Straub
Charles George Wegman

**1977**

Alexander Walter Bobiak
George Osei Bonsu
Raphael Bustin
John Hall Chase, Jr.
Alice P. Check
Thomas Franklin Check
Yuk Y. Cheng
Florian Patrick Eggleton
George Alan Fehlner
Gary Roger Fink
David D. French
Mark Alan Gohlke
John Kevin Grier
Gregory John Halliday
David Childs Jordan
Bruce Edmond Kelly
Robert Wallace Moote
Robert J. Pizzutiello
Daniel Adam Rapoport
Anthony Ravinsky
Peter Willy Reed
Alan Edward Roll
Martin David Rubin
Robert James Ryan
Khalid Sayood
Kenneth Edward Scheffter
Michael Shannon
Robert Laurence Stuhr
Mark Pullen Vandeusen
Christopher G. Winter

**1978**

Jonathan Cooper Allen
Becky Louise Badham
David A. Butler
King Fai Choi

Michael Dennis Christie
Frank Lawrence Clifton
Jeffrey Don Davis
Richard A. Druyeh
Michael R. Engbrecht
Benjamin Ross Epstein
Bruce Martin Epstein
Steven Lewis Feierstein
Jean Lynne Hearn
Jon Charles Hiller
Gordon Kenneth Kapes
George Vincent Kondraske
Wing Kong Law
William D. Lentz
Anil Narang
Jeffrey Dart Nathan
Ari M. Novis
David John Olek
James Richard Oslica
Duane Edward Reid
Janet Helen Robinson
Ronald Wayne Rotach
John Joseph Sabol, Jr.
Scott Andrew Sampi
Mark Edward Schmid
Douglas M. Schwarz
Jonathan Paul Shore
Janet Sterritt
Jackal Jeun-Cheung Tse
David Wayne Warker
Stanley Smith Wirsig
Iris Young-Sheppard
Stephanie Marie Zwolinski

**1979**

David Edwin Aiken
Anthony James Albanese

Carl Michael Blahut
Boleslaw F. Boczkaj
Anthony Richard Bonaccio
Richard Edward Burney
Steven Chi F. Chan
Tai Chun Danny Chow
Jeffrey Anthony Coriale
Bonnie Jean Couchman
Laurence A. Crosby
Paul Allan DeHart
Jodi Fox
Reazul Haque
Michael David Kaiser
William David Lentz
Khalid Mahboob
Steve Montellese
Ira Moskowitz
Earl Martin Norman
Ronald Padrov Novick
Musa Moiz Ovadya
Stephen Edward Paluszek
Kumaresan Satchithanandam
Eng Joo Sin
Janet E. Slocum
David Howard Weiner
Joseph Peter Wood
Peter Yick Fai Wu

**1980**

William James Beaty
Brian Christopher Bell
Edmund Paul Chin
Joseph Peter DiVincenzo
Benjamin Moses England, Jr.
Joseph R. Farage
David Richard Finamore
Gary Roger Greene
Rita Renea Jackson

Enoch Kang
Alan M. Kielar
Kevin Charles Klem
Jasper James Lachiana
Shung Keung Lam
Howard Steven Lazarus
Cynthia B. Levy
Neil David McKay
Richard Stephen Milner
Joseph Edward Misanin
Charles Babcock Nevins, Jr.
Philip T. Nicklaus
Koray Oguz
Rikki Razdan
Susan Jean Resnick
Howard Schafer
Walter Francis Schultz
Mark Andrew Struebel
Alan Jay Swartz
John Richard Trotter
Tierry Tzau
Michael Wang, Jr.
Robert Harvey Wilcox, Jr.
Warren Scott Wolfeld
Mark William Worthington

**1981**

Eli J. Behlok
Daniel J. Blumenthal
Robert B. Bowser
Merrill Northington Bradley
Richard Joseph Breen
Stewart Abraham Bresler
Darrel James Damon
Jeffrey Alan Devoyd
Margaret Clark Gartner
Barry Wayne Goldin
Gary R. Greene

Fred V. Guteri
Allen W. Hale
Michael Donald Henry
William Tien En Ho
Jon Cameron King
Michael David Klein
Alex Chichung Kot
Peter Tat Kwan Kwok
Alan Bruce Langdon
Charles Alan Linn
Deborah Anna Ludlum
William Patrick Mann
Raymond A. Morris
Dorothy Lee Nixon
Michael Benson Richards
William Edward Saltzstein
Bryan Edward Santor
Elliot Sobel
Andrew T.H. Soh
Robert Henry Sorensen
Burton Mahler Strauss, III
Timothy Szczerbinski
Robert Alan Vieira
Robert Joseph Wanamaker, Jr.
Martin James Zuber

**1982**

Duncan Banks Barbee
Kirk Howard Bingeman
Gregory Edward Cholmondeley
William Ming Chu
Danny E. Clabeaux
Elaine Leslie Cook
Thomas Michael Fredette
Richard A. Hjulstrom
Ralph Stephen Jahnige
David Bert Johnston
Leonard Edward Kay

David A. Kloss
Kendall Elizabeth Martin
Scott Rowell McCleneghan
Kwok Yeung Ng
Peter Stewart Paulson
Alexander Norman Peck
Michael Robert Rosato
Joseph Darryl Sawicki
David Alan Schechter
David Sroka
Jeffrey Scott Stone
Benjamin Moses Vogel
William S. Woods
William Jeffrey Zwolinski

**1983**

Jonathan Salem Balke
William L. Barnard
Patti Allyson Bernstein
Steven Bruce Birenbaum
Marc Steven Birnkrant
Jeffrey William Blain
Pier David Carroll
Simon Wansui Chan
Michael En Chen
Timmy Chow Keong Chin
Katherine McKenica Durdan
Gary Alan Emerle
Steve Yung-Chieh Fan
Anne S. Fischer
Robert Alan Franken
Alan Michael Goldman
James Frederick Haskin
John Hutson
James A. Ionata
Gregory Charles Issing
James Knight Johnson

Brian Douglas Kain
Douglas Michael Karp
Matthew Raycroft Kercher
Gunyong James Kim
Dwight Allen Klaus
Thomas Edward Kopley
Gregory William Kress
Karla Marie Kuzawinski
May Lam
John William Lockard
Edwardo Antonio Mantilla
George Douglas Mazezka
Herman McKenzie
Frederick Alfred Mueller
Brian Henry Nicholas
Jeffrey Victor Peck
Linda Beth Pollack
Geoffrey Rust Prior
Sam M. Romano, Jr.
Mark A. Rudd
Murray Edward Rudin
John Stephen Sakosky
Donald Edwin Schildkamp
Julie R. Shapiro
Eddie Kam Wah Siu
Andrew L. Smith
Mark Seth Stoller
Thomas Bergan Wall
Carl R. Wesolowski

**1984**

Remigio Jose Arteaga
Roland Yap Bairan
Jeffrey Mark Barr
Stephen Joseph Beltramini
Thomas John Buto
Thomas Richard Carducci
David F. Carlson

Lorina Fung Chan
Michael L. Charash
Joe Yul Cho
John Francis Conneely
Thomas Frank Crider
Jerome Michael Daniszewski
Poul Eldrup-Jorgensen
Richard Thomas Fischer
Yves Diego Fontayne
Loren Charles Fox
Philip Sterling Freedman
Robert George Gingher
Jonathan Mark Goldman
Scott Richard Hammond
Mary Alice Jones
Tony Roy Kaytes
Paul David Kottler
Kenneth Samuel Kreit
Andrew Leo Kutter
Jeffrey Leung
Philip Evans McKinley
Christine Mary Miyachi-Bohner
Lawrence Wai-Shing Mui
Ismet O. Mustecaplioglu
Victor Albert Noel
Surjana Albert Noel
Surjana P.S. Oey
Kathleen Mary Owczarski
David C. Parkins
Robert Michael Pedro
Michael Joseph Randall
Steven Redlich
Samuel Neil Rosenberg
James Clayton Sharp, III
Marc D. Siegel
Richard Edward Sweetman, Jr.
Theresa Poy-Hing Szeto
Richard Edmund Thorne
Theresa Ann Tuthill
Nicholas Martin Vanheel

Margaret Lin Veruki
Jeffrey Victor Wickenheiser
Kurt William Youngs
Michael Eugene Zugger

**1985**

Edward K. Au
John Curtis Austin
Leslie Elaine Battistoni
Mondher Ben-Ayed
Rachel Bess Berg
Khaled Bettaieb
Theresa Anne Boone
Julianne Boyd
Edward Stephen Broda
Paramjit Singh Chadha
Deborah Yvonne Chin
Soo Mae Chu
Pamela Louise Church
Matthew Barbour Dwyer
William G. Fonte
Eric G. Forrest
David L. Forster
David Stuart Freidin
Eric McKenzie Frietas
Mark Richard Gabay
Elliot Drew Garbus
Douglas W. Gardner
Charles Jeffrey Garrison
David William Geen
Kurt R. Hacke
Randall Keith Hems
Jay Scott Hersh
Stephen Hogan
Peter Tat Huie
Michael Ford Hunt
Moez Jedidi

Ridha Kamoua
Michael Lawrence Kaplan
Glenn Thomas Kincaid
Kevin Lee Kornher
Richard J. Krulik
Saher Lahouar
Danielle Beth Laibowitz
Lawrence Brian Landry
Henry Lee
Kip Rowland Leitner
Jonathan David Levy
Tod Eric Luginbuhl
Nabil Maalej
Mehrdad Mahamedi
Richard Thomas Mann
Peter T. Maricle
Carrick Alan McLaughlin
Andrew Stevenson Miller
Barbara Davis Molnar
Joseph Michael Muhitch
Deborah Z. Nifong
Thomas Joseph O'Shea, Jr.
Joseph Edward Oravetz
Eric F. Pelletier
Yinfun Poon
William J. Phillips
Dean Michael Rabbitt
Paschal John Romano
Neil A. Rosen
Brion Daryl Sarachan
Mark S. Scott
Tracy Leigh Shearing
Stephen Jon Siegel
Kevin Paul Smyntek
Carl Hudson Staelin
Laurence Jon Sternbane
Matthew Lawrence Taradash
Nicolas Tausanovitch
John B. Taves
Gimtong Teowee

Christopher Robert Teumer
Mark Edward Todd
James Weidman, IV
Matthew Brennan Weppner
Christopher Morris Wolff

**1986**

James Ronald Anuskiewicz
Amy Boardman Barnhart
Andrew John Bartfay-Szabo
Erik C. Becker
Anthony Borza
Eileen M. Brown
Harpaul Singh Chadha
Daniel Wing Hong Chan
Steven Tseng-Hung Chen
Ching-Yao Chu
Howard Hwa Chu
Lon Mae Chu
Christopher James Cicchetti
James Philip Coene
Gary Robert Crane
Christopher J. Daley-Watson
Gary Benjamin Depue
Peter John Deweerdt
Paul Craig Dorsey
Michael M. Falk
Mark R. Garber
Mark David Gartland
Jeremy Grodberg
David John Hinshaw
Nichola B. Hosack
Joel Andrew Hughes
Eric Henry Jones
Peter Simon Kashin
William Leung
Jason V. Macres
Charles J. Meyers

Michael Bernard Monahan
Robert Allison Moore, Jr.
Lorna Jeanne Morehead
John William Morris
Thomas James O'Day
Theodore Donald Ousback
John W. Parkins
Daniel Colin Perez
Gregory Pien
David Lee Porterfield
Clyde Wendell Robinson, Jr.
Gustave Anthony Ruth, Jr.
David Anthony Salman
Kevin Paul Siegel
William James Simonds
Edward Mark Stairman
Michael Gregory Steinthal
Robert Ragan Stephenson
Mark Kenji Takita
Kit-Ming Wendy Tang
Atilla Terzioglu
David Aron Tobias
Frederick Hoff Treesh
Bruce David Verner
Lisa Mei-Sheung Wan
Todd Andrew Waterman
Marc Warren Webster
Judith Ann Weiler
Charles Richard Wild

**1987**

David Robert Adams
Domenic John Alcaro
Jill C. Alfert
Stuart Ross Barich
Christopher Jon Bucci
Michael C. Chang

Okryong Chun
Carmen Chun-Kwong Cook
Christie Carol Cramer
Scott Thomas DeGhetto
Christopher Edward Eron
Julie Ann Fisher
Michael Aaron Fisher
Barbara Ann Fuegel
Geri Ilene Goldman
David Mark Gostin
Kenneth William Gray
Atul Kumar Gupta
Jae Won Kim
Constantine Kontanis
Laurie Conway Kornher
Andrew Edward Krause
Howard Allen Lazoff
Quang Xuan Le
Steven Landon Lew
Daniel Charles Luongo
Burt Magen
William Churchill Merrill
Sze-Him Ng
Stephen Andrew Olsen
Lloyd Frederick Palum
Kenneth Jerome Peters
Mitchell Jay Price
Drew Miles Randolph
Howard Eliot Reichel
Kirk Alan Reinbold
Peter Wallace Schimpf
Peter Daniel Sulatycke
Thomas Stephen Tirone
Thomas Nathaniel Tombs
Apostolos Tsoukkas
Walter F. Wafler
Alan Marshal Warwick
Marjorie Eleanor Weibel
Mathieu Wiepert
Mark Donald Wiltse

Eric Martin Ziegler
Kenneth Edmund Zweig

**1988**

James Michael Battiato
Charles Terence Jesse Bennington
Glen Erik Bohling
David Gerard Burkhardt
Catherine Heddy Carbonaro
Alfred Clinton Carr
David Andrew Casper
Stephen Boyce Culbert
Mario Robert DiRubbo
Duane Raymond Douglas
James William Duff
Laurina Anne Ferro
Gordon Scot Franza
Benjamin James Goldthwaite
Jeffrey Lynn Gorsuch
William Augustus Grazier
Jayne Catherine Hahin
Chehreh Harir
Christopher Peter Helgesen
Timothy W. Henkels
S. Hamilton Hitchings
Jerome Vincent Hochmuth
Mark Peter Hughes
Jon Martin Jacob
Karl David Karpa
Shahid Ahmad Khan
Dipinder S. Khurana
Gautham Krishnan
Randi H. Kutzin-Constant
Brenda Lee Luderman
Daniel Frederick Luna
Robert Mark Margolis
Michalakis Andronicos Michaelides
Theodore Edward Miller

Raymond S. Moyer
Rakesh Mullick
Yoshikazu Nagatsuka
Howard Christopher Read
Scott Edward Richardson
Nancy Scala
Eric Hartmut Schierling
Michael John Seymour
Timothy Jonathan Talley
Lee Samuel Wagmeister
Brent Jack Wahba
Bruce Robert Waring
Neil David Washburn
Mark Steven Weiner
Daniel A. Welch

**1989**

John Andrew Benoit
John Allen Carter
Ming-Hsin Michael Chang
Kaushik Chaudhuri
David T. Chen
Edward Chun-Chich Chen
Wai Shun Bruce Cheng
Richard Andrew Cirincione
Matthew Christopher Coriale
Michael W. Dailey
John Joseph Debole
Theodore Drapas
Vira Em
Kevin Eugene Gordon
Wai Wah Ho
David Daniel Katzoff
Daniel Joseph Langdon
Jonathan Sotha Lei
Peter Joseph Lynch
David Joseph Macannuco
Suzanne Pamper Marcellus

David Gabriel Miranda
David Gary Morrison
Paul Charles Nauert
Horacio Munsayac Padua, Jr.
Antonette Susan Pettinato
Ronald Matthew Quain
Jonathan Karl Riek
Johnny Ka-Lun Szeto
Karen Udy
David James Walters
Vivyan Lee Weinman
Tammy Sweeney Welder
John Edward Werner
Donna Marie White
Stephen Gareth Williams
Stuart Curtis Zak

**1990**

Wing Keung Frederick Au
Emre Munip Bilgi
George Stephen Blasiak
Patrick J. Borrelli
Darren Keith Brock
Walter Douglas Carl
Yueh-Shiun Chan
Philip Arnold Ferolito
John Iovieno
Razak Hossain
Chong Dae Kim
John Patrick King
Robert Alan Kocsmiersky
Paul Christopher Krein
Jane Chit Lam
Chung Fai Edmond Lee
Luke Liem
Jerry C. Lin
Robert Francis Marchese
Elaine Beth Meyer

Robert Michael Misita, Jr.
John David Quinzi
Kashif Ijaz Rana
Thomas Ristich
Robert E. Roskos, Jr.
David William Schuh
David James Scoggins
Yihchih Shern
Andrezej Sobski
Kenneth Ivan Steinberg
Jing Wang

**1991**

Anita T. Acre
Timothy John Arrowsmith
Melissa Ann Barber
Scott Barnes
Jeffrey Michael Blum
Eric Warren Brown
Tim Chiu
Hyun Cho
Thomas Leo Conroy, II
Muzaini Bin Darus
Brian Joseph Flynn
Tadd Bernard Hughes
Chia-Kan Hung
Jun Yong Jang
Lakshmi Kanthan
Michael David Kocsmiersky
John Alexander Krakuszeski
Mee-Mee Lai
Barry A. Lederman
John P.Y. Leng
Fai Li
Richard Lun
John Douglas MacKay
Paul Matthew Muntner

Frank Hans-Harald Norrmann
Steven Anthony O'Toole
Michael Robert Pantano
Deborah G. Peyton
David Schuchman
Richard Christopher Scinta
Susan F. Seyrek-Esen
Kevin Paul Stanton
Robert A. Todd, II
Tzong-Maw Tsai
Jonathan S. Ty
Amirali Vellani
Phillip Anthony Volz
Chen Kuang Yu

**1992**

Peter Leo Allen
Steven James Arrigenna
Mark Chamberlin Bahler
Paul Brian Benning
Alexander James Brashich
Matthew Eric Carlson
Allan Chi Yan Cheung
Wei-Wei Chew
Stephen Choy
Carol Ann Dann
Mitchell Steven Diamond
Daniel Lawrence Duperron
Martin Rolf Fick
Hei Tao Fung
Tony Leroy Green
Andrew King Halberstadt
Christine Regis Hilow
Stacy Michele Kahalas
Koon-Keung Lai
Christine Marjo Langton
Gerald Michael Lihota

Dennis Kevin Lockshine
Peter Theodore Mendick
Rebecca L. Millard
Stephen Joseph Muscato
Khai Thien Nguyen
Leonard Prokopets
Sujeeva Senaka Ranasinghe
Mohammad Athar Shah
Usha Shastry
Zita J. Sidas
Kevin A. Simpson
Harvinderpal Singh
Andrew Weldon Stonner
Robert Joseph Sturtz
Nicholas Leon Susch
Whitfield Robinson Thompson
Theodore Kendrew Tyree
Matthew William Warmuth
Jeffrey William Wilcox
Marshall Crawford Wilson
John M. Wortman
Masri Bin Yakop
David M. Zacher

**1993**

Daniel Berkowitz
Mark Andrew Bernard
Kara Gay Burgan
Robert T. Burns
David Matthew Bussard
Carl Frederick Conrath
Sean Antoine Drakes
Akiva Elijah Elias
Jose Sandoval Garcia
Nathan Geryk
Bryan Wayne Goettsch
David Scott Goldman

Navneet Govil
Mark David Hahm
Nor Hisham Hamid
Bryan Michael Hausner
Adriana Higuera
Dexter W. Hodge
Ike Aret Ikizyan
Anthony John Jamberdino
Brian Willis Jensen
Steven Kong
Chao Hsu Lee
Peter L. Malave
Wayne A. Miller
Roy Muermann
Joseph A. Nelli
Everett Gibson Newbry, III
Vincent Tuan Nguyen
Kelly Lynn Nowak
Christopher Alan Paczkowski
Peter James Parshall
Erik Michael Schlanger
Matthew Howard Schmukler
Matthew Howard Silver
Charles Alfred Streb, IV
Esin Terzioglu
Craig Carlton Truax
Hung Tsang
Jeffrey Alan Waldrop
David Lyndon Warnke
Linda Wong
Frances H. Yuan

**1994**

Keith Brian Blodorn
Orville Karram Brown
Andrew Scott Burchill
Errol Vinton Campbell
James John Carbonaro

Kevin Roy Carlson
Kevin Daniel Champaigne
Michael Justin Ciszek
Frantzy Duverne
Arjuna Ekanayake
John D. Gambon
Deepak Goyal
Judy Fong Huang
Richard Kirkcaldy Hynds
Robert Kaser
Shannon Colleen Kelly
Charles G. Lederman
Valentine Kong Chee Loh
Gregory John Lukins
Dhiraj Mallick
Abdul Halim Bin Haji Abdul Manaf
Brian Robert McGee
Lincoln Alphanso McLoud
Ann Sheridan Mihalik
Ali Muhtaroglu
Mahran Mohammed Najeeb
Meng Ni
Chukwuma Ejekam Nnolim
Darren Granger Nowell
Kevin M. Olson
William Hamilton Phipps
David Scott Reyner
Kevin Francis Rhatigan
Diane Santos
Anup Kamal Sharma
Cordigon Cornel Taylor
James Venetsanakos
William John Vanremmen
Daniel Francis Vona
Kenneth James Weaver
Carlo Anthony Williams
Brian Paul Willnecker
Di Wu
Ming Jian Wu
Fai Yeung

**1995**

Salman Moiz Ahmed
Frederick Anton Beer
Joseph Aaron Blatt
Gueorgui Dobrinov Borchoukov
Matthew Wyatt Buttrick
Jerrid Donald Chapman
William Andrew DeCanio
Sandeep Dudhwewala
Vipul C. Gandhi
Lior Gurwitz
Kevin William Harrison
Choochart Haruechaiyasak
Ling Cherd Ho
Christopher Bingham Johnson
Murat Berkan Kacar
Aaron David Kneiss
Irene M. Krenskaya
Peter Hun-Ding Leung
Chris Alan Long
Khurram Zaka Malik
Basab Mukherjee
Philip Kyalo Mutooni
Dmitry Netis
Daniel Anthony Nosek
William David Oliver
Shakeel H. Peera
Chhem Kim Pen
Mathew William Pettis
Scott Andrew Rasche
Damir Saracevic
Ahmed Shahid
Beejal Kantilal Shah
Radha Shreeniwas
Marc Douglas Spencer
Pornchai Supnithi
Hikmet Emre Ucok
Christopher Harold Van Leuvan
Robin Alicia Weinberg

Jeffrey Randall Wilcox
Zhe Zheng

**1996**

Nauman Afzal
Tahseen Ahsan
Jeet Sudhir Asher
Hasnat Mahmood Ashiq
Rakhee Bhattacharya
Kenyon P. Binns
Benjamin David Gannon
Jeffrey Michael Gramowski
Jason McDonald Green
Manosh John
Sandesh S. Kaveripatnam
Khizar Ahmad Khan
Alisa Kongthon
Evan Louis Kurtz
Otto Cheung Pui Lam
Jason Alexander Leonard
Stephen Andrew McAleavey
Collin Orlando McCreath
Lalit Kumar Mehta
Faisal Mohammed Mir
Roy Nolen Mounier
Anne Marie Murphy
Sardar Faisal Nisar
Michael J. Orr
Rahul Vijay Panday
Thomas Richard Parker
Rebecca L. Peterson
Brian Christopher Porter
Steven James Pratt
Zaid Safdar
Nalinda Sapukotana
Saurabh Pratap Singh
Chubasco C. Spivey
Pranjal Srivastava

Gennousuke Takahashi
William John Vanremmen
James Theodore Viahos
Chau-Jye Yang
Andy Wu
Wei Zhu

**1997**

Christopher David Agnew
Kerstin Elaine Babbitt
Anthony Dominick Boccio
Michael Chad Bristol
Carlos Nathan Catanach
Wendy Wen Ying Chiu
Tanzeem Khalid Choudhury
Agnim I. Cole
Kerry J. Denvir
Shiloh Leigh Dockstader
Terrance M. Jones
Ashish Kumar
Sidney Christopher Laurenceau
Kaushik Mittra
Sumit Mohan
Benjamin Munger
Kin Ng
John Francis Palma
Daniel Jacob Petrovich
Matthew A. Reh
C.M. Nabeel Sami
Philip Paul Schremmer
Asif Iqbal Shah Mohammed
Zhi-Jian Shen
Daniel E. Stephens
Edwin Sulaiman
Abbas Tahir
Chi Kin Tang
Akshat Ajay Thanawala
Jonathan David Tushman

Justin Eric Vlietstra
Gu Feng Yin
Ren-Yu Zhang

**1998**

Marchelle Christine Adams
Adrian Robert Cancel
Hyouk-Jean Cha
F.B. Amitha Buddhika De Silva Weeras
Rahul Gupta
Nicholas Andrew Hirth
Yosuke Inoue
Hye-On Jang
Michelle Shi-Ying Lo
Micaiah Jay McTaggart
Jason Prunty
David Allen Roberts
Michael David Rodgers
Gregorio Ciardi Sanchez
Gregory Scott Sullivan
Din-Yi Sun
Syed Mohammad Ali Tahir

**1999**

Benjamin Stanislaw Cichy
Leonardo Cordaro
Jeffrey Paul Davis
Derek Brendan Gottlieb
Adam Gerald Joslyn
Scott Alan Knauss
Zamsury Kushaili
Aaron Steven Master
Matthew Sidare
Mario T. Simpson
Randolph N. Singh
John Thomas Strassner

Scott David Teitsch
Alexandra Torres
Allyn J. Turner
Urso Atila Vargas
Frank Michael Verdico
Chao Wang

# Notes

## Notes to Chapter 1

1 Except in the office of the University Registrar, student records are no longer accessible within the university community since The Family Educational Rights and Privacy Act (FERP) was passed by the U.S. Congress in 1974.
2 Jesse L. Rosenberger, *Rochester: The Making of a University* (Chicago: University of Chicago Press, 1927); Arthur J. May, *A History of the University of Rochester, 1850–1962* (Princeton, N.J.: Princeton University Press, 1977); John C. Friedly, *75 years of Chemical Engineering at the University of Rochester, 1915–1990* (Rochester: University of Rochester Publication, 1990).

## Notes to Chapter 2

1 Percy Dunsheath, *A History of Electrical Power Engineering* (Cambridge, Mass.: MIT Press, 1962), 103; Charles Susskind, "American Contributions to Electronics: Coming of Age," *Proceedings of the Institute of Electrical and Electronic Engineers (IEEE)* 64:9 (September 1976): 1300.
2 *IEEE Transactions on Education*, E-23 (August 1980): 169.
3 *The Forty-Fifth Catalog of the Officers and Students of the University of Rochester, 1894–1895*, 60.
4 *Annual Catalog of the University of Rochester, 1904–1905*, 75.
5 *Report of the President of the University of Rochester, 1905–1906*, 9.
6 Jesse L. Rosenberger, *Rochester: The Making of a University* (Rochester: University of Rochester, 1927), 280.
7 *Report of the President of the University of Rochester, 1908–1909*, University of Rochester archives, 3.
8 Ibid., 9.
9 Ibid., 4.
10 *Annual Catalog of the University of Rochester, 1909–1910*, 27–28.
11 More information on the early days of engineering at the university can be found in chapter 1 of John C. Friedly, *75 years of Chemical Engineering at the University of Rochester, 1915–1990* (Rochester: University of Rochester, 1990).

12 J.H. Belknap report, January 1948, Belknap Public Relations file, University of Rochester archives.
13 University of Rochester News Release, August 20, 1945, Public Relations file, University of Rochester archives.

## Notes to Chapter 3

1 Report of the Dean, College of Engineering for 1961–1962, University of Rochester archives.
2 University of Rochester News Release, February 21, 1958, University of Rochester archives.
3 E.L. Carstensen, D.W. Healy, Jr., H.P. Schwan, and S. Talbot, "Curriculum Development in Biomedical Engineering," *J. Eng. Education* 53 (1963): 446.
4 *IRE Transactions on Medical Electronics*, ME-7 (October 1960): 255.
5 *IRE Transactions on Medical Electronics*, ME-7 (January 1960): 31.
6 *IRE Transactions on Circuit Theory* 8 (September 1961): 237.
7 Report of the Dean, College of Engineering and Applied Sciences for 1969–1970, University of Rochester archives, 28.

## Notes to Chapter 4

1 E. Titlebaum and R.A. Altes, "Bat Signals as Optimally Doppler Tolerant Waveforms," *Journal of the Acoustical Society of America* 48 (1970): 1014.
2 Society of Women Engineers, *Statistics About Women in Engineering in the U.S.A.*, http://www.swe.org, November 15, 2002.
3 P. Bowron and F.W. Stephenson, *Active Filters for Communications and Instrumentation* (London: McGraw-Hill, 1979).
4 Charles W. Merriam, *Automated Design of Control Systems* (New York: Gordon and Breach, 1974); and *Solutions to Optimization Problems Arising in Feedback Control with Fortran Computer Program* (Lexington, Mass.: Lexington Books, 1978).
5 Lloyd Hunter, ed., *Handbook of Semiconductor Electronics*, 3rd ed. (New York: McGraw-Hill, 1979).

## Notes to Chapter 5

1 H.C. Flynn and C.C. Church, "Transient Pulsations of Small Gas Bubbles in Water," *J. Acoust. Soc. Am.* 84 (1988): 1863.
2 Edwin L. Carstensen, *Biological Effects of Transmission Line Fields* (New York: Elsevier, 1987).

## Notes to Chapter 6

1 *Science Watch* (October 1991): 7.
2 Thomas B. Jones, *The Electromechanics of Particles* (Cambridge, Mass.: Cambridge University Press, 1995).
3 A. Murat Tekalp, *Digital Video Processing* (Garden City, N.J.: Prentice-Hall, 1995).
4 Eby G. Friedman, ed., *High Performance Clock Distribution Networks* (Boston: Kluwer Academic Publishers, 1997).
5 Alan M. Kadin, *Introduction to Superconducting Circuits* (New York: Wiley Interscience, 1999).

## Notes to Chapter 7

1 From Huston Smith, "Two Kinds of Teaching," in T.H. Buxton and K.W. Prichard, eds., *Excellence in University Teaching: New Essays* (Columbia: University of South Carolina, 1975), 207.

# Name Index

Abdallah, Mahmoud, 133
Abrams, R., 31
Abrishamkar, Farrokh, 132
Abushagur, Mustafa A.G., 76
Ackerman, Eugene, 31
Acre, Anita T., 170
Adam, Roman, 142
Adams, David Robert, 165
Adams, Marchelle Christine, 178
Adams, Mark, 71
Adkin, John V., 18, 145
Adler, Victor Seth, 122, 139
Afzal, Nauman, 176
Agajanian, Aram, 152
Agarwala, J.S., 129
Agnew, Christopher David, 140, 177
Ahmed, Salman Moiz, 175
Ahsan, Tahseen, 176
Aiken, David Edwin, 155
Ajewole, Isaac Ishola, 112
Akselrod, Gennady L., 141
Alam, Sheikh Kaisar, 120, 138
Albanese, Anthony James, 155
Alberti-Merri, Michela, 137
Albicki, Alexander, 61
Albonesi, David, 85
Alcaro, Domenic John, 165
Alexander, Bruce Allen, 136
Alexandrou, Sotiris, 118, 137
Alfert, Jill C., 165
Allen, Carlton C., 20
Allen, Jonathan Cooper, 154
Allen, Peter Leo, 171
Allyn, Elwyn G., 28, 146
Alphenaar, Arthur W., 28, 125
Alte, Charles Roger, 149
Altes, Richard A., 49, 111, 127
Altunbasak, Yucel, 121, 139
Anderson, Eric C., 149
Anderson, Martin, 94
Anderson, Robert Frank, 150
Ang, Marcelo Huibonhoa, Jr, 76, 115, 135
Angel, Edward, 54
Ansaldi, Silvia, 77
Anstee, Richard Paul, 147
Anthony, Donald, 153
Antonini, Eleazar E. James, 150
Anuskiewicz, James Ronald, 136, 164
Appleby, Robert Alan, 138
Apte, Dilip Madhav, 131
Araghi, Mehdi N., 112
Arakere, Prathima, 129
Archer, Stephen John, 135
Arden, Bruce W., 71
Arellano-Ramirez, Manuel Enrique, 129, 168
Argento, Joseph J., 149
Armstrong, Del, 77
Armstrong, Gavin, 56
Armstrong, Michael F., 29, 146
Aronstein, Robert H., 28, 126
Arora, Rajiv, 117, 134
Arrigenna, Steven James, 171
Arrowsmith, Timothy John, 170
Arteaga, Remigio Jose, 160
Asami, Shigeo, 39
Asher, Jeet Sudhir, 176
Ashiq, Hasnat Mahmood, 176
Ashton, Edward Andrew, 120, 139

Ashton, James Joseph, 29, 146
Astheimer, Jeffrey P., 77
Ateya, Antoun I., 133
Attarian, Farshid, 142
Au, Edward K., 162
Au, Wing Keung Frederick, 169
Ault, Wilbur E., 18, 145
Auman, William Daniel, 133
Aurtenechea, Francoise, 140
Austin, John Curtis, 162
Avari, Nausheer Jal, 129
Ayme, Eveline Jeannine, 115, 134

Babbitt, Kerstin Elaine, 177
Bacchetta, Richard William, 148
Backus, Lee F., 28, 146
Bacon, David, 49
Bacon, Howard Elston, 12
Badham, Becky Louise, 154
Badilini, Fabio Francesco, 118, 138
Bahler, Mark Chamberlin, 171
Baig, Taqi M., 142
Bailat, Julien, 98
Bailey, Jack Frederick, Jr, 151
Bairan, Roland Yap, 160
Baird, John, 17
Bakonis, William Leo, 151
Balke, Jonathan Salem, 159
Balke, Sheila Z., 152
Balke, Thomas R., 152
Ballentine, Paul Henry, 119
Ballentine, Paul W., 135
Balling, Alfred, 14
Balzer, David Mark, 132, 153
Bancrof, Christopher Marier, 150
Bannon, James K., 127
Bara, Zygmund J., 18, 145
Barbee, Duncan Banks, 158
Barber, Melissa Ann, 170
Barich, Stuart Ross, 165
Barmish, B. Ross, 55
Barnard, William L., 159
Barnes, Scott, 170
Barnhart, Amy Boardman, 164
Baron, Cynthia Jean, 137
Barr, Jeffrey Mark, 160
Barry, Thomas William, 151
Bartfay-Szabo, Andrew John, 164
Bartlett, Robert, 134
Basehore, Richard Michael, 151
Basnet, Bishwa Vijaya, 138
Basu, Asish, 69
Battiato, James Michael, 167
Battistoni, Leslie Elaine, 162
Bayer, Gerald W., 56
Beattie, Larry William, 135
Beaty, William James, 156
Beausang, James Anthony, 116, 134
Bechtold, Robert, 14
Becker, Erik C., 164
Becker, Robin D., 153
Becroft, Steven Allen, 132
Beebe, William Dunning, 150
Beer, Frederick Anton, 175
Behlok, Eli J., 157
Belknap, John Harrison, 14
Bell, Brian Christopher, 156
Bell, Kevin Richard, 151
Bellegarda, Jerome Rene, 116, 134
Beltramini, Stephen Joseph, 160
Ben-Ayed, Mondher, 91, 116, 135, 162
Bender, John Frederick, 153
Bennett, James C., 152
Benning, Paul Brian, 171
Bennington, Charles Terence Jesse, 167
Benoit, John Andrew, 168
Benwood, Bruce Robert, 130
Berg, Rachel Bess, 162
Bergman, Steven Charles, 129
Berkcan, Ertugrul, 115
Berkowitz, Daniel, 172
Bernard, Mark Andrew, 172
Bernstein, Alan, Jr, 29, 147
Bernstein, Patti Allyson, 159
Berson, Bertrand E., 29, 126
Bessenyei, Bela, 130
Bettaieb, Khaled, 162
Betz, John Wells, 153
Bhatia, Sandeep, 137
Bhattacharya, Ameet S., 141
Bhattacharya, Rakhee, 176
Bhattacharyya, Amalendu B., 76
Bhella, Kenneth S., 140

Bilgi, Emre Munip, 169
Bilheimer, Robert, 69
Bingeman, Kirk Howard, 158
Binns, Kenyon P., 176
Birenbaum, Steven Bruce, 159
Birnkrant, Marc Steven, 159
Birrell, N. Kirk, 132
Bishop, Peter B., 150
Blackstock, David T., 31, 76, 96
Blahut, Carl Michael, 156
Blain, Jeffrey William, 159
Blasiak, George Stephen, 169
Blatt, Joseph Aaron, 175
Blezurs, Brigita, 77
Blodorn, Keith Brian, 173
Blum, Jeffrey Michael, 170
Blumenthal, Daniel J., 157
Bobiak, Alexander Walter, 154
Boccio, Anthony Dominick, 177
Bocko, Mark F., 64
Boczkaj, Boleslaw F., 156
Bodmann, Joseph J., Jr, 147
Bogardus, Russell Carl, 130
Bohling, Glen Erik, 167
Bonaccio, Anthony Richard, 156
Bonino, Paul Samuel, 142
Bonsu, George Osei, 154
Boone, Theresa Anne, 162
Boonpikum, Ittibhoom, 141
Boonyanant, Phakphoom, 141
Booth, Irwin S., Jr, 18, 145
Borchoukov, Gueorgui Dobrinov, 175
Bordoni, Franco, 76
Borrelli, Patrick J., 140, 169
Borza, Anthony, 164
Bossert, Clement O., 18, 145
Bowman, Charles K., 28, 146
Bowman, Robert, 73
Bowser, Robert B., 157
Boyd, Julianne, 162
Boynton, Robert, 30
Boyse, John M., 56
Boyton, Robert M., 35
Bozdagi, Gozde, 98
Bradley, Merrill Northington, 157
Brainard, Douglas C., 152
Brashich, Alexander James, 171
Breen, Richard Joseph, 157
Bresler, Stewart Abraham, 157
Bristol, Michael Chad, 177
Brock, Darren Keith, 121, 140, 169
Broda, Edward Stephen, 162
Brown, Brian D., 138
Brown, Edward S., 18, 145
Brown, Eileen M., 164
Brown, Eric Warren, 170
Brown, George Arthur, 28, 125
Brown, James, 69
Brown, Orville Karram, 173
Brown, Ralph J., 18, 145
Brusil, Paul John, 149
Bucci, Christopher Jon, 137, 165
Buchiere, Philip J., 18, 145
Budd, Gary Kean, 131
Burattini, Laura, 121, 140
Burchill, Andrew Scott, 173
Burgan, Kara Gay, 172
Burkhardt, David Gerard, 167
Burney, Richard Edward, 156
Burns, Robert T., 172
Burns, Stephen, 69
Burstein, William Jerry, 149
Burton, Henry F., 10
Bussard, David Matthew, 172
Bustin, Raphael, 154
Butler, David Alan, 133, 154
Butler, Donald Philip, 115, 133
Butler, Zeynep Çelik, 116
Buto, Thomas John, 160
Buttrick, Matthew Wyatt, 175

Cain, Charles, 71
Cambier, James Lacey, 55, 113, 130
Campbell, Derrick Shawn, 137
Campbell, Errol Vinton, 173
Campbell, James Andrews, 114, 132
Campbell, Lloyd Reginald, Jr, 28, 146
Cancel, Adrian Robert, 178
Cannon, William John, 147
Cao, Wei, 136
Caravaglio, Francis J., 28, 146
Carbonaro, Catherine Heddy, 167
Carbonaro, James John, 173

Carducci, Thomas Richard, 135, 160
Carey, Graham, 77
Carl, Walter Douglas, 169
Carlson, David Frederic, 135, 160
Carlson, Frederic Roy, Jr, 148
Carlson, John R., 152
Carlson, Kevin Roy, 174
Carlson, Matthew Eric, 171
Carr, Alfred Clinton, 167
Carroll, Pier David, 159
Carstensen, Edwin L., 25, 107
Carter, John Allen, 168
Case, Timothy John, 139
Casper, David Andrew, 167
Catanach, Carlos Nathan, 177
Celasum, Isil, 98
Çelik, Zeynep, 116
Celly, Nikhil, 138
Cetin, Enis, 76
Cha, Hyouk-Jean, 178
Chadha, Harpaul Singh, 164
Chadha, Paramjit Singh, 162
Chakravarty, Shanti P., 111, 128
Champaigne, Kevin Daniel, 174
Chan, Daniel Wing Hong, 136, 164
Chan, Kah-Fae, 131
Chan, Lorina Fung, 161
Chan, Patrick S., 133
Chan, Selena, 142
Chan, Simon Wansui, 159
Chan, Stephen Chi Fai, 115, 134, 156
Chan, Yueh-Shiun, 169
Chana, Maghar Singh, 130
Chang, Michael C., 165
Chang, Michael Ming Hsin, 120, 138, 168
Chang, Yi-Hua Edward, 129
Chaolin, 62
Chapman, Jerrid Donald, 175
Chapman, William A., 151
Charash, Michael L., 161
Chase, John Hall, Jr, 154
Chaudhuri, Kaushik, 168
Cheah, Chin Hong, 141
Check, Alice P., 154
Check, Thomas Franklin, 56, 154
Chen, Chang Wen, 96
Chen, Chiou-Shiun, 110, 126, 128
Chen, David T., 168
Chen, Edward Chun-Chich, 168
Chen, Lei, 143
Chen, Lulin, 77
Chen, Michael En, 159
Chen, Steven Tseng-Hung, 164
Chen, Xucai, 89
Chen, Yuk Yuen, 132
Cheng, Wai Shun Bruce, 168
Cheng, Wing Kai, 152
Cheng, Yuk Y., 154
Cherkauer, Brian S., 119, 138
Cheung, Allan Chi Wan, 138, 171
Chew, Wei-Wei, 171
Chiang, Edward S.I., 29, 126
Chik, Lawrence L., 55, 110, 128
Chikte, Shirish, 55
Child, Sally Zehr, 36, 94
Chin, Deborah Yvonne, 162
Chin, Edmund Paul, 156
Chin, Timmy Chow Keong, 159
Chiu, Tim, 170
Chiu, Wendy Wen Ying, 177
Chivers, Robert C., 76
Cho, Hyun, 170
Cho, Joe Yul, 135, 161
Choi, King Fai, 154
Choice, Lawrence E., 148
Chokchaitam, Somchart, 140
Cholmondeley, Gregory Edward, 158
Choudhury, Tanzeem Khalid, 177
Chow, Tai Chun Danny, 156
Choy, Stephen, 171
Christie, Michael Dennis, 155
Christopher, Paul Edward Ted, 69, 118, 134
Chu, Ching-Yao, 164
Chu, Howard Hwa, 164
Chu, Lon Mae, 164
Chu, Soo Mae, 162
Chu, William Ming, 158
Chuang, Shih-Shung, 111, 128
Chun, Okryong, 166
Chung, Ji-Yong David, 136
Church, Charles 68
Church, Pamela Louise, 162
Chwalek, James Michael, 135

Cicchetti, Christopher James, 164
Cichy, Benjamin Stanislaw, 178
Cidale, David R., 150
Ciesielski, Maciej J., 114
Cirincione, Richard Andrew, 168
Ciszek, Michael Justin, 174
Clabeaux, Danny E., 158
Claflin, Gerald E., 29, 147
Clark, Alan L., 77
Clark, Charles Henry, Jr, 152
Clark, Guy Peter, 128
Clark, Marvin Dale, 28, 146
Clark, Wesley, 30
Clarke, W. Bromley, 28, 110, 126
Clayton, Richard, 56
Cleveland, Bruce M., 129
Clifton, Frank Lawrence, 155
Cliver, Richard C., 140
Clynes, Manfred, 30
Coene, James Philip, 136, 164
Cohen, Gerald Howard, 23
Cohen, Mark Alan, 151
Cole, Agnim I., 143, 177
Coleman, Andrew, 71
Coleman, P., 31
Coleman, Thomas G., 28, 146
Combs, Cecil E., 40
Compton, John Thomas, 139
Conneely, John Francis, 161
Connin, John Lyman, 128
Conrath, Carl Frederick, 172
Conroy, Thomas Leo, 170
Conta, Lewis D., 18
Converse, Frederick James, 12
Cook, Carmen Chun-Kwong, 166
Cook, E.R. Gordon, 55
Cook, Elaine Leslie, 158
Coombs, William, 23
Cordaro, Leonardo, 178
Coriale, Jeffrey Anthony, 156
Coriale, Matthew Christopher, 168
Corner, Thomas C., 151
Correia-Neves, Jose Luis P., 138
Couchman, Bonnie Jean, 156
Coulton, Derek A., 150
Cowan, Charles Rodgers, 150
Cox, Christopher, 69
Cramblitt, Robert M., 98
Cramer, Christie Carol, 166
Crandall, Guy O., 14
Crane, Gary Robert, 164
Crawford, Thomas J., 133
Crider, Thomas Frank, 161
Crockett, Abraham, 71
Crosby, Laurence A., 156
Cross, John William, Jr, 148
Cruikshank, Donald B., Jr, 111, 127
Crum, Larry, 70
Culbert, Stephen Boyce, 167
Currie, Marc Daniel, 122, 139
Curry, Robert C., 28, 125
Curtis, Johnson Ottawa, 148
Czernikowski, Roy S., 65

Dailey, Michael W., 168
Dakin, R. Edward, 153
Dalecki, Diane, 69, 88, 94, 118, 134
Daley-Watson, Christopher J., 164
Daley, Michael L., 112
Damon, Darrel James, 157
Daniszewski, Jerome Michael, 161
Dann, Carol Ann, 171
Dans, Ronald Frank, 148
Darus, Muzaini Bin, 170
Das, Pankaj K., 34
David, Edward, 31
Davis, Albert Tatum, 117
Davis, Jeffrey Don, 155
Davis, Jeffrey Paul, 178
Davis, Miles, 28, 125
Davis, Richard Murray, 147
Davy, Bruce A., 149
Dawson, Charles Holcomb, 13
Day, James Edward, 148
De Kiewiet, Cornelius W., 21
Debole, John Joseph, 168
DeCanio, William Andrew, 175
DeGhetto, Scott Thomas, 166
DeHart, Paul Allan, 156
Deihl, David Thompson, 148
Delesky, Walter W., 152
Deletto, Thomas Anthony, 131
Delfyett, Peter John, Jr, 134

Delgado, Jose Luis, 129
Delisio, John P., 152
Delius, Michael, 71
Demczar, Barbara F., 136
Denk, Tracy C., 138
Dennis, William J., Jr, 152
Denvir, Kerry J., 177
Denysenko, Andrew, 139
Depue, Gary Benjamin, 164
Derefinko, Victor V., 55, 75
DeRoller, Charles L., 77
Devoyd, Jeffrey Alan, 157
Dewcerdt, Peter John, 164
Dhanesha, Hema Mohanial, 139
Diamond, Mitchell Steven, 171
Dijak, Jerome Thomas, 150
DiRubbo, Mario Robert, 167
Dise, Donald P., 18, 145
DiVincenzo, Joseph Peter, 156
Dmitriev, Vytaly, 41
Dobrolis, Konstantinos M., 141
Dockstader, Shiloh Leigh, 142, 177
Dodsworth, John, 77
Donaldson, William, 65
Dorsey, Paul Craig, 164
Dosani, Azim Alibhai Velji, 131
Doughty, Thomas E., 18, 145
Douglas, Duane Raymond, 167
Douglas, James Livingston, 25, 125
Douglas, McDonald, 50
Douglas, Ruth, 116, 135
Draeger, Carsten, 98
Drakes, Sean Antoine, 172
Drapas, Theodore, 168
Druyeh, Richard A., 155
Dubash, Noshir Behli, 119, 137
Dudek, Stanley J., 28, 125
Dudhwewala, Sandeep, 175
Duerr, Jeffrey R., 147
Duff, James William, 167
Dumic, Marc J., 76
Dunlap, John Hallowell, 149
Duperron, Daniel Lawrence, 171
Durdan, Katherine McKenica, 159
Dutcher, Barry Carl, 17
Duverne, Frantzy, 174
Dwarkadas, Sandhya, 85
Dwyer, Matthew Barbour, 162
Dykaar, Douglas Raymond, 115, 133

Eames, Frederick Allyn, 131
Eberly, Joseph H., 41
Ebner, Fritz Francis, 137
Eccles, Kathleen Marie, 153
Eckel, Vincent Louis, 152
Eckhardt, Norman Lester, 153
Efron, Adam Joshua, 132
Egerton, Ian, 98
Eggleton, Florian Patrick, 154
Ehrich, Roger W., 147
Einolf, Charles William, Jr, 110, 127
Eisen, Jane Kira, 135
Ekanayake, Arjuna, 174
Elberfeld, John Karl, 129
Eldrup-Jorgensen, Poul, 161
Eleiott, Charles, 135
Elias, Akiva Elijah, 172
Eliasson, Ingvar E., 18, 145
Elkind, Steven Alan, 153
Eller, Anthony Irving, 29, 109, 126
Em, Vira, 168
Emerle, Gary Alan, 159
Emiris, Dimitrios, 117, 136
Emrick, Daniel Herbert, 131
Enders, Timothy M., 137
Engbrecht, Michael R., 155
England, Benjamin Moses, Jr, 156
Epstein, Benjamin Ross, 155
Epstein, Bruce Martin, 155
Erdem, Arif Tanju, 74, 116, 136
Eren, Pekin Erhan, 141
Ernsberger, Millard Clayton, 12
Eron, Christopher Edward, 166
Erturk, Erdal, 71
Evanitsky, Eugene Stephen, 132
Evans, William C., 126
Everbach, E. Carr, 69, 76, 97
Ewing, Joan Rose, 51, 112, 128

Falk, Michael M., 164
Faller, David G., 138

Fan, Steve Yung-Chieh, 159
Fang, Yu, 140
Farage, Joseph R., 156
Farber, David J., 76
Farden, David C., 54
Fasoli, James Steward, Jr, 153
Fauchet, Philippe M., 84, 92, 107
Fehlner, George Alan, 154
Feierstein, Steven Lewis, 155
Feinberg, Alan B., 151
Feldman, Jerome A., 46
Feldman, John, 149
Feldman, Marc, 64
Fenton, Daniel Bruce, 151
Ferman, Ahmet Mufit, 142
Ferolito, Philip Arnold, 169
Ferro, Laurina Anne, 167
Fick, Martin Rolf, 171
Fields, David, 54
Finamore, David Richard, 156
Fink, Gary Roger, 154
Firth, Paul Bruce, 152
Fischer, Anne S., 159
Fischer, Richard Thomas, 161
Fisher, Julie Ann, 166
Fisher, Michael Aaron, 118, 137, 166
Fisher, Wendell Burns, Jr, 132, 151
Flint, Orin Queal, Jr, 129
Flood, Robert Francis, Jr, 151
Flores, Assis Miranda, 152
Flynn, Brian Joseph, 170
Flynn, Hugh Guthrie, 24, 40
Fontayne, Yves Diego, 161
Fonte, William G., 162
Fork, David Kirtland, 136
Forman, Ernest H., 147
Formaniak, Peter G., 130
Forrest, Eric G., 162
Forster, David L., 162
Fowler, James E., 135
Fox, Jodi, 156
Fox, Loren Charles, 161
Fradet, Erwan, 98
Francis, Charles, 89
Frank, John, 18, 146
Frankel, Michael Yuri, 136
Franken, Robert Alan, 159
Franza, Gordon Scot, 167
Frazo, Arthur, 54
Fredette, Thomas Michael, 158
Freed, Gerald Lewis, 28, 146
Freedman, Philip Sterling, 161
Freidin, David Stuart, 162
French, David D., 154
Friedly, John C., 59
Friedman, Eby G., 82, 107
Friedman, Leonard, 150
Frietas, Eric McKenzie, 162
Frizzell, Leon A., 113, 130
Fry, William, 31
Fu, Yue, 142
Fuegel, Barbara Ann, 166
Fujita, Hiroyuki, 137
Fuller, Lynn, 65
Fullmer, David Merlin, 135
Fung, Hei Tao, 120, 139, 171
Furchill, Patrick Anthony, 142

Gabay, Mark Richard, 162
Gafni, Giora, 111
Gaj, Krzysztof, 91
Galimidi, Alberto Ricardo, 134
Galombus, R., 31
Gambon, John D., 174
Gammons, Richard Allan, 135
Gamo, Hideya, 34
Gandhi, Vipul C., 175
Gannon, Benjamin David, 176
Gao, Lan, 119, 140
Garber, Mark R., 164
Garbus, Elliot Drew, 162
Garcia, Jose Sandoval, 172
Gardner, Douglas W., 162
Gardner, Kenneth, 14
Gardner, Marvin, 69
Garlick, Richard Alan, 128
Garrison, Charles Jeffrey, 162
Gartland, Mark David, 164
Gartner, Margaret Clark, 157
Gaspar, Alfred Frederick, 126
Gavett, Joseph W., Jr, 12
Geen, David William, 162
Geiger, Steven Paul, 151

Geller, Eric Mitchell, 131
George, Nicholas, 95
Geryk, Nathan, 172
Ghosh, Dipankar, 133
Giesselmann, Albert Carl, 18, 145
Gilbert, Steven Galen, 151
Gimian, Gideon, 131
Gingher, Robert George, 161
Givens, Miles Parker, 71
Glogowski, Leo T., 147
Gob, Wolfgang, 98
Goding, Justin Christian, Jr, 116, 132
Goettsch, Bryan Wayne, 172
Gogowski, Leo Thomas, 127
Gohlke, Mark Alan, 154
Goins, James Allen, Jr, 151
Goldak, John, 77
Goldin, Barry Wayne, 157
Goldman, Alan Michael, 159
Goldman, David Scott, 172
Goldman, Geri Ilene, 166
Goldman, Jonathan Mark, 161
Goldman, Stanford, 30
Goldstein, Fred T., 56
Goldstein, Julius L., 35, 109
Goldstein, Moise, 30
Goldthwaite, Benjamin James, 167
Goltsman, Grigori, 96
Gong, Ting, 98
Goodman, Lawrence, 30
Goodrich, Lewis Miner, 18, 145
Gordon, Douglas, 36
Gordon, Kevin Eugene, 168
Gordy, Edwin, 31
Gorsuch, Jeffrey Lynn, 167
Gostin, David Mark, 166
Gottlieb, Derek Brendan, 178
Govil, Navneet, 173
Goyal, Deepak, 174
Graber, Christopher C.H., 147
Grace, Robert Edmond, 111
Gracewski, Sheryl, 69
Graeper, William G., 18, 146
Graham, John W., Jr, 20, 21
Graham, Philip Allan, 149
Gram, Martha, 69
Gramiak, Raymond, 36
Gramowski, Jeffrey Michael, 176
Gratian, J. Warren, 29, 126
Gray, Daniel Alan, 135
Gray, J.F., 30
Gray, Kenneth William, 166
Gray, Martin Bennett, 127
Grazier, William Augustus, 167
Green, Jason McDonald, 176
Green, Kenton Andrew, 122, 140
Green, Tony Leroy, 171
Greene, Elliott S., 151
Greene, Gary Roger, 156
Grier, John Kevin, 154
Grodberg, Jeremy, 164
Grodins, Fred, 30
Guan, Bo Ran, 98
Gunes, Dogan, 152
Gunsel, Bilge, 98
Gupta, Anshoo Sudhir, 129
Gupta, Atul Kumar, 166
Gupta, Deepnarayan, 120, 139
Gupta, Rahul, 178
Gupta, Shantanu, 136
Gupta, Siddhartha Prakash Dutta, 121
Gurwitz, Lior, 175
Gustavson, Dann Alan, 151
Guteri, Fred V., 158
Gwaltney, Mark A., 141

Haavik, Stanley Jens, 127
Habif, Jonathan, 87
Hacke, Kurt R., 162
Hager, Stephen Mark, 151
Hah, Zaegyoo, 96
Hahin, Jayne Catherine, 167
Hahm, Mark David, 121, 139, 173
Halberstadt, Andrew King, 139, 171
Hale, Allen W., 158
Hallemeier, Peter F., 140
Haller, William Joseph, 131
Halliday, Gregory John, 154
Hamid, Nor Hisham, 173
Hamilton, Clark Allen, 35, 111, 128
Hammond, Scott Richard, 161
Haque, Reazul, 156
Harir, Chehreh, 167

Harp, Ellen Veronica, 139
Harper, Henry Roland, 150
Harris, Matt John, 135
Harris, Thomas Jerome, 25
Harrison, Kevin William, 175
Hartquist, E. Eugene, 50, 149
Haruechaiyasak, Choochart, 175
Haskin, James Frederick, 159
Hattersley, Thomas E., Jr, 28, 126
Hausner, Bryan Michael, 173
Hayes, David F., 150
Hayes, Peter Rodney, 151
Hayman, Scott M., 135
Hayter, Alan Beckwith, 151
Headlam, David, 84
Healey, Daniel J., 147
Healy, Daniel Ward, 22
Hearn, Jean Lynne, 155
Heath, Robert Raymond, 127
Hegmann, Frank, 96
Heinzelman, Wendi, 101
Heiter, Christian G.M., 137
Helgesen, Christopher Peter, 167
Helguera, Maria, 138
Helmers, Peter Halfdan, 153
Hems, Randall Keith, 162
Henderson, Kenneth C., 141
Hendrickson, Roy Hulme, 12
Henkels, Timothy W., 167
Henry, Michael Donald, 158
Hercher, Michael, 52
Hernandez, Robert Lee, 149
Herr, Quentin Paul, 121
Herrmann, James Francis, 138
Hersh, Jay Scott, 162
Herzog, Gilbert W., 128
Heurtley, John Crampton, 109
Hewitt, William Joseph, 149
Heywood, Timothy Charles, 141
Higuera, Adriana, 173
Hiller, Jon Charles, 155
Hilow, Christine Regis, 171
Hinshaw, David John, 164
Hirsch, Victor J., 151
Hirth, Nicholas Andrew, 178
Hitchings, S. Hamilton, 167
Hjulstrom, Richard A., 158
Ho, Ling Cherd, 175
Ho, Wai Wah, 168
Ho, William Tien En, 158
Hoang, Viet-Dzung, 151
Hochmuth, Jerome Vincent, 167
Hodge, Dexter W., 173
Hoefer, Robert Joseph, 18, 145
Hoffmann, Peter, 56
Hoffmeister, J. Edmund, 16
Hogan, David Charles, 130
Hogan, Stephen, 162
Hoie, Erling, 98
Hollot, Christopher Valentine, 76, 114, 133
Holmes, James William, 136
Hon, Henry Hengwah, 152
Honig, Carl, 30
Hopeman, Albert A., 20
Hopeman, Arendt, 20
Hopeman, Bertram C., 26
Hopkins, Dana Brand, Jr, 147
Hopkins, Mark R., 77
Hopkins, Robert E., 41
Hornung, Christoph, 77
Horowitz, Deborah Ronnie, 133
Horwitz, James W., 76
Hosack, Nichola B., 164
Hossain, Razak, 119, 139, 169
Houde, Robert A., 127
Houkes, Fulco, 98
Hsiang, Thomas Y., 61
Hsieh, Chao-Peng, 126
Huang, Judy Fong, 174
Huang, Kenneth Y, 152
Huang, Sung-Rung, 116, 136
Huang, Sung, 98
Huang, Ter-Tsu, 132
Hughes, Joel Andrew, 164
Hughes, Mark Peter, 167
Hughes, Tadd Bernard, 170
Huie, Joseph A., 28, 125
Huie, Peter Tat, 162
Huitric, Herve, 77
Hung, Chia-Kan, 170
Hunt, Michael Ford, 162
Hunter, Lloyd P., 34, 107
Hurt, Philip Francis, 132

Hussong, Frederick Alan, 149
Hutchison, Bruce Raymond, 132, 153
Hutson, John, 159
Hynds, Richard Kirkcaldy, 141, 174

Ikizyan, Ike Aret, 173
Iline, Konstantin, 96
Imaide, Takuya, 133
Inchalik, Michael Allen, 135, 137
Indaco, Paul Vincent, 131
Infortuna, Ann H., 136
Ingkutanon, Non, 141
Ingram, Marylou, 41
Inoue, Yosuke, 178
Ionata, James A., 159
Iovieno, John, 169
Isemann, Robert Clayton, 18, 146
Ismail, Yehea Ismail, 142
Issing, Gregory Charles, 159
Iwakami, Etsuo, 132

Jackson, Rita Renea, 156
Jackson, Thomas Humphrey, 82
Jackson, Todd A., 115, 133
Jacob, Jon Martin, 167
Jacobs-Perkins, Douglas W., 122, 139
Jacobs, John, 31
Jahnige, Ralph Stephen, 158
Jain, Anil Kumar, 111, 129
Jamberdino, Anthony John, 173
Jammy, Rajarao, 139
Jang, Hye-On, 178
Jang, Jun Yong, 170
Jansson, Tomas, 98
Jarnjak, Stjepan A., 131
Jedidi, Moez, 162
Jensen, Brian Willis, 173
Jigno, Han, 77
John, E. Roy, 30
John, Manosh, 176
Johnson, Christopher Bingham, 175
Johnson, James Knight, 159
Johnson, Mark, 98
Johnson, Richard Alan, 112, 128
Johnson, Ronald Alvin, 149
Johnson, Roy Clifford, 17
Johnson, Willard C., 152
Johnston, David Bert, 158
Jones, Eric Henry, 164
Jones, Mary Alice, 161
Jones, Michael S., 152
Jones, R., 30
Jones, Robert Hawthorne, 90, 149
Jones, Terrance M., 177
Jones, Thomas B., 73, 107
Jones, Thomas D., Jr, 148
Jones, Warren R., 152
Jordan, David Childs, 154
Joshi, Dilip Pandurang, 130
Joslyn, Adam Gerald, 178
Jovancevic, Aleksandar Velimir, 108, 122, 141
Jungo, Marc, 98

Kacar, Murat Berkan, 175
Kacprzak, Tomasz, 76
Kadin, Alan M., 64
Kahalas, Stacy Michele, 171
Kain, Brian Douglas, 160
Kaira, Devendra, 132
Kaiser, Michael David, 156
Kaiser, Richard F., 18, 146
Kaler, Karan V., 76
Kalinowski, Jerzy, 135
Kalmanash, Michael H., 127
Kamak, Stephen Glen, 147
Kamath, Venkatesh H., 130
Kamoua, Ridha, 163
Kanda, Ryoichi, 139
Kane, Peter, 77
Kang, Enoch, 157
Kanthan, Lakshmi, 170
Kantrowitz, Arthur W., 26
Kapadia, Varsha, 135
Kapes, Gordon Kenneth, 155
Kaplan, Michael Lawrence, 163
Karim, Munawar, 77
Karnik, Avinash Ramkrishna, 110
Karp, Douglas Michael, 160

Karpa, Karl David, 167
Karsky, Michael K., 28, 126
Kaser, Robert, 174
Kashin, Peter Simon, 164
Kassabian, Dikran, 95, 139
Katsadas, Evagelos, 117, 136
Katzoff, David Daniel, 168
Kaveripatnam, Sandesh S., 176
Kay, Leonard Edward, 158
Kaytes, Tony Roy, 161
Ke, Qing, 119, 139
Kedem, Gershon, 50
Keenan, John E., 127
Keenan, Thomas A., 45
Kegelman, Arthur Scott, 149
Kehoe, Michael P., 153
Kellock, David E., 129
Kelly, Bruce Edmond, 154
Kelly, Shannon Colleen, 174
Kennedy, Robert Phelps, Jr, 18, 145
Kercher, Matthew Raycroft, 160
Kern, Marana Anne, 132
Kettig, Robert Lawrence, 149
Khalid, Zaim, 139
Khan, Khizar Ahmad, 176
Khan, Shahid Ahmad, 167
Khanna, Dev Karan, 129
Khare, Manoj, 136
Khoo, Audrey Mei-Ling, 135
Khromenko, Constantin, 41
Khurana, Dipinder S., 167
Kicherer, William Carl, 148
Kielar, Alan M., 157
Kiker, William E., 72
Kim, Chong Dae, 169
Kim, Gunyong James, 160
Kim, Jae Won, 166
Kim, Man Bae, 98
Kincaid, Glenn Thomas, 163
King, James George, 148
King, John Patrick, 169
King, Jon Cameron, 158
King, Ronald Wayne, 148
Kingslake, Rudolf, 52
Kinnen, Edwin, 34, 82
Kinney, Daniel Steven, 133
Klaus, Dwight Allen, 160
Klein, Lawrence Arnold, 128
Klein, Lawrence Eliot, 19
Klein, Michael David, 158
Kleiner, Edward H., 129
Klem, Kevin Charles, 157
Klinger, Lance T., 147
Kloss, David A., 159
Knauss, Scott Alan, 178
Kneiss, Aaron David, 175
Knodt, Kurt Thomas, 138
Knox, Keith Thomas, 150
Kobyashi, Tetsuo, 77
Kocsmiersky, Michael David, 170
Kocsmiersky, Robert Alan, 169
Kohli, Ujjaldip Singh, 133
Koide, George Takashi, 110
Kondraske, George Vincent, 155
Kong, Steven, 173
Kongthon, Alisa, 176
Kontanis, Constantine, 166
Kopley, Thomas Edward, 160
Kornher, Kevin Lee, 163
Kornher, Laurie Conway, 166
Kortkamp, Krista Linn, 138
Kosalathip, Voravit, 141
Kostic, Zoran Ilija, 117, 136
Kostusiak, Karl H., 127
Kot, Alex Chichung, 158
Kottler, Paul David, 161
Kourtev, Ivan Stefanov, 123, 141
Koyama, Hideki, 98
Kozminski, Krzysztof Antoni, 114
Krakuszeski, John Alexander, 170
Krasniewski, Andrzej, 87, 135
Krause, Andrew Edward, 166
Kraybill, John P., 135
Krein, Paul Christopher, 169
Kreit, Kenneth Samuel, 161
Kremkau, Frederick William, 36, 112, 129
Krenskaya, Irene M., 175
Kress, Gregory William, 160
Krishnan, Gautham, 167
Kriss, Michael Allen, 95
Krol, Tamara G., 141
Kula, Witold, 96

Kumar, Ashish, 177
Kumar, Rakesh, 112
Kumer, K.N., 136
Kung, Sun-Yuan, 131
Kung, Thomas Ling-Wen, 151
Kurtz, Clark Nelson, 110
Kurtz, Evan Louis, 176
Kushaili, Zamsury, 178
Kushel, Vladimir, 28, 126
Kutter, Andrew Leo, 161
Kutzin-Constant, Randi H., 167
Kuzawinski, Karla Marie, 160
Kwok, Peter Tat Kwan, 158

Lacefield, James, 98
Lachiana, Jasper James, 157
Lagler, Regis Gary, 131
Lahouar, Saher, 163
Lai, Koon-Keung, 171
Lai, Kwok-Woon, 153
Lai, Mee-Mee, 139, 170
Lai, Weiwen, 141
Laibowitz, Danielle Beth, 163
Lakomy, Dale G., 132
Lam, Jane Chit, 169
Lam, May, 160
Lam, Otto Cheung Pui, 176
Lam, Shung Keung, 157
Landry, Lawrence Brian, 163
Lang, Wolfgang, 97
Langdon, Alan Bruce, 158
Langdon, Daniel Joseph, 168
Langton, Christine Marjo, 171
Lanzillotta, Joseph, 132
Lapp, Theodore Raymond, 149
Laskin, Alan Perry, 149
Laszio, Ernest J., 128
Lau, Randall Sek-Tim, 153
Laurance, Neal, 77
Laurenceau, Sidney Christopher, 143, 177
Law, Wing Kong, 155
Lawrence, Henry E., 10
Lazar, Nicholas, 18, 146
Lazarus, Howard Steven, 157
Lazoff, Howard Allen, 166
Le, Quang Xuan, 166
LeBlanc, Thomas, 82
Lederman, Barry A., 170
Lederman, Charles G., 174
Lee, Chao Hsu, 173
Lee, Chung Fai Edmond, 169
Lee, Henry, 163
Lee, Paul P., 55, 113, 131
Lee, Robert A., 56
Lee, Robert E., 109, 126
Lefor, John, 68
Lehman, Justus, 31
Lei, Jonathan Sotha, 168
Leinberg, Eric John, 136
Leitner, Kip Rowland, 163
Lemkin, Peter Freemman, 128
Leng, John P.Y., 170
Lentz, William David, 155
Leonard, Jason Alexander, 176
LePage, Wilbur Reed, 13, 22
Lerner, Robert Marc, 71, 113
Lessen, Martin, 90
Lettrin, Jerry, 30
Leung, Jeffrey, 161
Leung, Peter Hun-Ding, 175
Leung, Pok Ming, 131
Leung, William, 164
Levinsen, Mogens, 51
Levinson, Stephen Fred, 97
Levy, Cynthia B., 157
Levy, Jonathan David, 163
Lew, Steven Landon, 166
Lewin, John Morton, 135
Li, Fai, 170
Li, Haotao, 142
Liem, Luke, 169
Lihota, Gerald Michael, 171
Likharev, Konstantin K., 51
Lillie, Jeffrey S., 140
Lin, Feng, 142
Lin, Hou-Sheng, 141
Lin, Jerry C., 169
Lin, Yu-Hwan, 113
Lind, Paul Ulrich, 147
Lindgren, Sven Gunnar Mikael, 96
Linke, Charles, 48
Linn, Charles Alan, 136, 158
Lipatov, Andrey, 97

Liptak, Gregory, 69
Littlefield, Bruce Godley, 151
Liu, Bin, 118
Liu, Dong-Lai, 97
Liu, Xun, 142
Liu, Yu-Jih, 130
Llavina, Rafael, Jr, 128
Lo, Michelle Shi-Ying, 143, 178
Lockard, John William, 134, 160
Lockshine, Dennis Kevin, 172
Loewy, Robert G., 40
Loh, Valentine Kong Chee, 174
Lokanathan, Badri, 116, 135
Long, Chris Alan, 175
Longacre, Andrew, Jr, 112, 128
Lorenzo, Luis, 143
Lorenzo, Sharon Anne, 139
Lowenfeld, I., 30
Lowy, Karl, 31
Lubanko, David Mathias, 153
Lubberts, Gerrit, 112, 128
Lubin, Moshe, 47
Luderman, Brenda Lee, 119, 137, 167
Ludlum, Deborah Anna, 158
Luening, James Wallace, 128
Luginbuhl, Tod Eric, 163
Lugo, Eduardo, 98
Lui, Dong-Lai, 98
Lukins, Gregory John, 174
Lun, Debbie Chaofang, 139
Lun, Richard, 170
Luna, Daniel Frederick, 167
Luo, Jiebo, 120, 140
Luongo, Daniel Charles, 166
Lusted, Lee B., 29
Luthra, Vijay, 131
Lynch, Peter Joseph, 168
Lyons, Mark Edmond, 136

Ma, Francis Kwokyiu, 131
Maalej, Nabil, 163
Macannuco, David Joseph, 168
MacKay, John Douglas, 170
MacKay, Stuart, 31
Macres, Jason V., 164
MacRobbie, Alan G., 141
Magen, Burt, 166
Magyar-Wilczek, Lorand W., 28, 125
Mahamedi, Mehrdad, 163
Mahboob, Khalid, 156
Maher, James Clinton, 133
Maier, Gregory J., 147
Malave, Peter L., 173
Maley, Alfred John, 147
Malik, Khurram Zaka, 142, 175
Mallick, Dhiraj, 174
Mallorg, Dorek, 88
Mallory, Derek Scott, 120, 137
Manaf, Abdul Halim Bin Haji Abdul, 174
Mancini, Cesar Augusto, 123, 139
Mandel, Leonard, 41
Mandelbaum, Richard, 97
Manfredi, Albert Emil, 150
Maniloff, Jack, 41
Mann, Richard Thomas, 163
Mann, William Patrick, 158
Mantilla, Eduardo Antonio, 159
Marcellus, Suzanne Pamper, 168
Marchese, Linda Elizabeth, 121, 139
Marchese, Robert Francis, 169
Marchetti, Jay D., 136
Margala, Martin, 101
Margolis, Robert Mark, 167
Maric, Svetislav Vojislav, 117, 136
Maricle, Peter T., 163
Marisa, Richard J., 50
Maronian, Roupen H., 131
Marquis, Robert, 36
Marshall, Christopher Ian, 136
Marshall, John Douglas, 153
Martens, Alexander E., 127
Martin, Gregory James, 134
Martin, Kendall Elizabeth, 159
Martinet, Stephen Swanston, 120, 137
Maruggi, Edward, 77
Mast, T. Douglas, 98
Master, Aaron Steven, 178
Masyga, Jon David, 136
Mathiason, Sharon Ann, 134
Matsumoto, Goro, 56
Matteson, Ronald G., 28, 125
May, Arthur J., 17, 19

Maye, Bonnie Ann, 133
Mayer, Robert, 69
Mazezka, George Douglas, 160
McAleavey, Stephen Andrew, 142, 176
McClaine, John Lucian, 132
McClellan, Robert William, 134
McCleneghan, Scott Rowell, 159
McCourt, Michael A., 153
McCreath, Collin Orlando, 176
McCrory, Robert L., Jr, 66
McFee, Richard, 30
McGee, Brian Robert, 174
McKay, Neil David, 157
McKechney, William J., 29, 147
McKenzie, Herman, 160
McKinley, Philip Evans, 161
McLaughlin, Carrick Alan, 163
McLoud, Lincoln Alphanso, 174
McNeill, John Arthur, 138
McTaggart, Micaiah Jay, 178
Meckling, H., 46
Mehlig, Theodore B., 28, 126
Mehta, Lalit Kumar, 176
Mehta, Sanjay Kumar, 117, 136
Mehta, Sushrut, 98
Meisel, Gary D., 152
Mendick, Peter Theodore, 172
Merle, Clor William, 28, 125
Meron, Peretz, 76
Merri, Mario, 116, 136
Merriam, Charles W., 47, 107
Merrill, William Churchill, 166
Meteyer, Gregory John, 151
Metz, Henry S., 41
Metzger, Jeffrey, 56, 153
Meyer, Elaine Beth, 169
Meyers, Charles J., 164
Miceli, Christopher Matthew, 138
Michaelides, Michalakis Andronicos, 167
Michaels, Elise Marie Raffan, 143
Michaels, Thomas Benjamin, 130
Michelson, Arnold Mark, 128
Middleditch, Alan E., 56
Mihalik, Ann Sheridan, 174
Millard, Rebecca L., 172
Miller, Andrew Stevenson, 163
Miller, Donald L., Jr, 148
Miller, Morton, 48
Miller, Ronald A., 18, 145
Miller, Ruth Douglas, 116
Miller, Theodore Edward, 167
Miller, Wayne A., 173
Milley, Nicholas A., 29, 147
Milne, John F., 147
Milner, Richard Stephen, 157
Minor, James, 76, 148
Minor, Oscar E., 26
Mir, Faisal Mohammed, 176
Miranda, David Gabriel, 169
Misanin, Joseph Edward, 157
Misic, Vladimir, 124, 143
Misita, Robert Michael, Jr, 170
Mitchell, Johanna May, 142
Mitrovski, Zoran, 108, 122, 140
Mitsa, Theophano, 75, 117, 136
Mittra, Kaushik, 177
Mix, Ron, 69
Miyachi-Bohner, Christine Mary, 161
Mohammed, Asif Iqbal Shah, 177
Mohan, Sumit, 177
Moin, Arthur David, 150
Molari, Pier Gabriele, 76
Molnar, Barbara Davis, 163
Molnar, Stanley, 30
Molyneaux, Robert Francis, 117
Monahan, Michael Bernard, 165
Montellese, Steve, 156
Montes, Laurent, 98
Montonye, J. Terrence, 128
Moore, Duncan, 83
Moore, Robert Allison, Jr, 165
Moote, Robert Wallace, 154
Morehead, Lorna Jeanne, 165
Morris, John William, 165
Morris, Raymond A., 158
Morris, Thomas Wilde, 148
Morrison, David Gary, 169
Morse, Theodore H., 28, 146
Moskowitz, Ira, 156
Moss, Arthur, 30
Mottley, Jack G., 73
Mounier, Roy Nolen, 176
Mourou, Gerard, 62
Moxley, Richard, 74

Moyer, Raymond S., 168
Mueller, Edwin Phillip, 148
Mueller, Frederick Alfred, 160
Muermann, Roy, 173
Muhitch, Joseph Michael, 163
Muhtaroglu, Ali, 174
Mui, Lawrence Wai-Shing, 161
Muir, Tom, 49
Mukherjee, Basab, 141, 175
Mukhopadhyay, Amitabha, 134
Mullick, Rakesh, 168
Mullick, S.K., 76
Mundie, J., 30
Munger, Benjamin, 177
Muntner, Paul Matthew, 170
Murphy, Anne Marie, 176
Murthy, Sankaran Narayana, 113, 130
Muscato, Stephen Joseph, 140, 172
Mustecaplioglu, Ismet O., 161
Mutooni, Philip Kyalo, 175
Myklebust, Joel Bruce, 130

Nagata, D. Tetsuya, 97
Nagatsuka, Yoshikazu, 168
Nahas, Monique, 77
Najeeb, Mahran Mohammed, 174
Nakano, Hiroyuki, 137
Narang, Anil, 155
Nathan, Amos, 41
Nathan, Jeffrey Dart, 155
Nau, Dana S., 77
Nauert, Paul Charles, 169
Naus, David Alan, 127
Nauth, Peter, 77
Neelakantan, Kumar, 120, 136
Nelli, Joseph A., 173
Nelson, David Elmer, 113
Nelson, Thomas, 56
Nenadic, Nenad, 142
Netis, Dmitry, 175
Neves, Jose Luis Pontes Correia, 120
Nevins, Charles Babcock, Jr, 157
Newbry, Everett Gibson, III, 173
Newman, Charles W., 111, 127
Newton, David Bruce, 148
Ng, Kin, 144, 177
Ng, Kwok Yeung, 159
Ng, Sze-Him, 166
Ngai, Eugene Cheung-Chun, 153
Nguyen, Khai Thien, 172
Nguyen, Minh Nhat, 137
Nguyen, Vincent Tuan, 173
Ni, Meng, 174
Nichles, Harry R., 18
Nicholas, Brian Henry, 160
Nicklaus, Philip T., 157
Nickles, Harry R., 18, 145
Nifong, Deborah Z., 163
Nilsson, Jan Ove, 56
Ning, Ruola, 89
Nisar, Sardar Faisal, 176
Nixon, Dorothy Lee, 158
Niznik, Carol Ann, 130, 149
Nnolim, Chukwuma Ejekam, 174
Noel, Surjana Albert, 161
Noel, Victor Albert, 161
Norman, Earl Martin, 156
Norrmann, Frank Hans-Harald, 171
Nosek, Daniel Anthony, 175
Notovitz, William David, 134
Novick, Ronald Padrov, 156
Novis, Ari M., 155
Nowak, Kelly Lynn, 173
Nowell, Darren Granger, 174
Nuñez-Regueiro, José Ernesto, 122, 140
Nyborg, Wesley, 31
Nyhof, William G., 18, 146

O'Brien, Bernard, 14
O'Brien, George Dennis, 61, 71
O'Day, Thomas James, 165
O'Donnell, Michael Gerald, 129
O'Shea, Thomas Joseph, Jr, 163
O'Toole, Steven Anthony, 171
Oestreich, Clifford P., 28, 125
Oey, Surjana P.S., 161
Oguz, Koray, 157
Olek, David John, 155
Oliver, William David, 175
Olsen, Stephen Andrew, 166
Olson, Kevin M., 174
Onofrio, Roberto, 77

Oravetz, Joseph Edward, 163
Orban, Julius, 151
Orfitelli, William A., 135
Orr, Michael J., 176
Osadciw, Lisa Ann, 123
Osborne, Paul, 42
Osborne, Robert Frederick, 29, 126
Oslica, James Richard, 155
Ostrander, Lee E., 29, 109, 126
Ousback, Theodore Donald, 165
Ovadya, Musa Moiz, 156
Owczarski, Kathleen Mary, 161
Ozkan, Mehmet Kemal, 118, 137
Ozone, Koho, 127

Paccard, Jacques, 130
Paczkowski, Christopher Alan, 173
Padua, Horacio Munsayac, Jr, 169
Paige, Arlie E., 17
Pakin, Sait Kubilay, 143
Palacios, Angel, 137
Palamar, Peter T., 131
Palma, John Francis, 177
Palum, Lloyd Frederick, 166
Paluszek, Stephen Edward, 156
Pance, Aleksandar, 118, 137
Pance, Gordana, 119, 137
Panday, Rahul Vijay, 176
Pant, Bhaskarrao V., 151
Pantano, Michael Robert, 171
Park, Kenneth Kun-E, 112, 129
Park, Roswell, 31
Parker, Kevin J., 61, 82, 92
Parker, Martin, 135
Parker, Thomas Richard, 176
Parkhill, Douglas, 31
Parkins, David C., 77, 161
Parkins, John W., 165
Parks, Gregory Johnson, 151
Parratt, Steffen W., 136
Parry, William Robert, 128
Parshall, Peter James, 173
Paruchuri, Venkatanarayana, 134
Patten, Stanley F., 41
Patti, Andrew John, 120, 138
Paul, Wray E., 141
Paulson, Peter Stewart, 159
Pavlovic, Gordana Miroslav, 118, 136
Paxton, K. Bradley, 111, 128
Payne, R. Alan, 148
Peale, Franklin, 69
Peck, Alexander Norman, 159
Peck, Donald George, 149
Peck, Jeffrey Victor, 160
Pedro, Robert Michael, 161
Peera, Shakeel H., 175
Pelletier, Eric F., 163
Pen, Chhem Kim, 175
Peng, Cheng, 120, 140
Penney, David, 69
Pentland, Alexander, 95
Pentland, Alice, 95
Perez, Daniel Colin, 165
Persson, Hans W., 56
Perucchio, Renato, 66
Perusko, Uros, 56
Pessel, David, 54
Peters, Kenneth Jerome, 166
Petersen, Ian Richard, 114, 133
Peterson, Bruce Alan, 132
Peterson, Donald P., 56
Peterson, L., 30
Peterson, Rebecca L., 176
Petronio, Julie, 97
Petrovich, Daniel Jacob, 177
Pettinato, Antonette Susan, 169
Pettis, Mathew William, 175
Peyton, Deborah G., 171
Phear, Robert C., 147
Phillip, Steve Glenville, 142
Phillips, Daniel Brian, 122, 140
Phillips, Joseph, 18, 145
Phillips, Robert G., 149
Phillips, William J., 163
Phipps, Arthur Raymond, 28, 126
Phipps, William Hamilton, 174
Pidel, Jeffrey Stephen, 132
Pien, Gregory, 165
Pilgrim, John George, 137
Pilkington, Wayne Charles, 143
Pillman, Bruce H., 139
Pizziconi, Louis F., 116
Pizzutiello, Robert James, Jr, 132, 154

Platnick, David, 28, 109, 126
Pohlig, Stephen C., 152
Policano, Thomas Joseph, 149
Pollack, Linda Beth, 160
Pollard, Robert Q., 17, 125, 145
Pollock, Elliott I., 18, 145
Poon, Yinfun, 163
Poppenberg, Richard Herman, 148
Porter, Brian Christopher, 142, 176
Porterfield, David Lee, 165
Portnoy, Jeffrey H., 150
Pratt, Steven James, 142, 176
Pratt, Thomas L., 137
Presher, Gordon Eugene, Jr, 149
Price, Mitchell Jay, 166
Principe, Arthur R., 18, 146
Prior, Geoffrey Rust, 160
Proctor, Thomas, 28, 126
Prokopets, Leonard, 172
Prunty, Jason, 178
Putney, William Matthews, 152

Quain, Ronald Matthew, 169
Quinn, James Michael, 153
Quinzi, John David, 170

Rabbitt, Michael, 163
Rahman, Timothy J., 148
Raji, Tesleem Iyola, 113
Rajpal, Ashok Jaikishen, 133
Rakoski, Frank Peter, 150
Rana, Kashif Ijaz, 170
Ranasinghe, Sujeeva Senaka, 172
Randall, Michael Joseph, 161
Randolph, Drew Miles, 166
Randolph, Walter J., 18, 145
Rao, Karnam Rameswar, 128
Rao, Kodati Subba, 29, 126
Raphael, Alan Frank, 151
Rapoport, Daniel Adam, 154
Rasche, Scott Andrew, 175
Ravinsky, Anthony, 154
Razdan, Rikki, 157
Read, Howard Christopher, 168
Reardon, Timothy John, 137
Redlich, Steven, 161
Reed, Peter Willy, 154
Regensburger, Paul Jerome, 136
Reh, Matthew A., 177
Reichel, Howard Eliot, 166
Reid, Duane Edward, 155
Reid, Jack, 31
Reid, Michael M., 34
Reilly, Clarence Raymond, 137
Reinagel, Frederick G., 28, 126
Reinbold, Kirk Alan, 166
Rella, Angelo Frank, 150
Renbeck, Robert Buch, 127
Requicha, Aristides A.G., 50, 66, 111, 128
Resnick, Susan Jean, 157
Reyner, David Scott, 174
Reyner, Noel L., 127
Rhatigan, Kevin Francis, 174
Rhees, Benjamin Rush, 10
Rhoads, Willard J., 130
Rhode, Robert Glen, 141
Rhodes, Richard Olney, 130
Rich, Neil Eric, 132
Richards, Michael Benson, 158
Richardson, Scott Edward, 168
Ridder, Jan, 56
Riek, Jonathan Karl, 119, 137, 169
Riess, Edward A., 150
Riethmeier, Alton F., 129
Rigby, Ronald Elwood, 150
Rink, Richard Andrew, 112, 128
Risley, Curtis Allen, 148
Ristich, Thomas, 170
Roberts, David Allen, 178
Robinson, Clyde Wendell, Jr, 165
Robinson, David, 30
Robinson, Janet Helen, 155
Rockefeller, 104
Rodgers, Michael David, 178
Rogers, Raymond J., 28, 126
Roll, Alan Edward, 154
Romano, Paschal John, 163
Romano, Sam M., Jr, 160
Rommelmann, Heiko, 134
Roopchansingh, Vinai, 142
Rosa, Michael, 134

Rosato, Michael Robert, 159
Rose, Paul James, 134
Rosen, Neil A., 163
Rosenberg, Samuel Neil, 161
Rosenberger, Jesse L., 11
Rosenblum, Kenneth Neil, 151
Roskos, Robert E., Jr, 170
Rosner, Richard Alan, 129
Rossignac, Jaroslaw Roman, 115, 134
Rotach, Ronald Wayne, 155
Roychowdhury, Vwani P., 134
Rubega, Robert Anthony, 28, 109, 126
Rubens, Deborah, 94
Rubin, Martin David, 154
Ruchkin, Daniel S., 25
Ruckdeschel, Frederick R., 112, 130
Rudd, Mark A., 160
Rudin, Murray Edward, 160
Rudolph, Alfred Robert, 149
Rugo, Tanya Lynel, 137
Rulli, Paul A., 133
Rummer, David Bruce, 151
Runyon, James C., 127, 129
Russell, Jeffery Mikko, 141
Russo, David Alan, 153
Ruth, Gustave Anthony, Jr, 165
Ryan, John Edward, 150
Ryan, Robert James, 154

Saber, Eli F., 139
Saber, Eli Said, 97, 120
Sabol, John Joseph, Jr, 155
Sadighi, Farhad, 153
Safdar, Zaid, 176
Sahmel, Rainer H., 148
Saia, Anthony, 77
Saini, Vasant Durgadas, 55, 114, 131
Sakosky, John Stephen, 160
Salata, Michael F., Jr, 128
Sales, Tasso, 98
Salman, David Anthony, 165
Saltz, Peter Alan, 129
Saltzstein, William Edward, 158
Sami, C.M. Nabeel, 177
Sampi, Scott Andrew, 155
Samuel, Nuriel, 56
Sanchez, Gregorio Ciardi, 178
Sanderson, Robert Lawrence, 110, 127
Sanger, Kurt M., 140
Santor, Bryan Edward, 158
Santos, Diane, 174
Sapukotana, Nalinda, 176
Saracevic, Damir, 175
Sarachan, Brion Daryl, 163
Sarna, Lalit Saran, 142
Satchithanandam, Kumaresan, 156
Saunders, Jeffrie Warren, 149
Sawicki, Joseph Darryl, 159
Saxena, Rajesh, 129
Sayood, Khalid, 133, 154
Sayuk, Keith Andrew, 153
Scacchetti, Armando, 29, 147
Scala, Nancy, 168
Schafer, Howard, 157
Schechter, David Alan, 159
Scheda, Mark Robert, 135
Scheffter, Kenneth Edward, 154
Schenk, Eric, 69
Schertler, Karl Egon, 130
Schierling, Eric Hartmut, 168
Schildkamp, Donald Edwin, 160
Schimpf, Peter Wallace, 166
Schlanger, Erik Michael, 173
Schmid, Mark Edward, 155
Schmukler, Matthew Howard, 173
Schnacky, Bernard J., 18, 145
Schremmer, Philip Paul, 177
Schrock, Anthony Ward, 140
Schuchman, David, 171
Schuh, David William, 170
Schultz, Walter Francis, 157
Schwan, Herman, 25
Schwartz, Gerhardt, 30
Schwarz, Douglas Mitchell, 133, 155
Schwarz, Karl Q., 89
Scinta, Richard Christopher, 171
Scinta, Ronald Christopher, 139
Scinta, Wendy Marie, 140
Scoggins, David James, 170
Scott, Cecil E., 18, 145
Scott, Mark S., 163
Seah, David I., 139
Secareanu, Radu Mircea, 142

Selwyn, Steven David, 150
Semenov, Alexey, 97
Seraphim, Samuel, 56
Serareanu, Radu M., 82
Seymour, Michael John, 168
Seyrek-Esen, Susan F., 171
Sezan, M. Ibrahim, 97
Shaftel, Myles Arthur, 131
Shah, Beejal Kantilal, 175
Shah, Himanshu Bhogilal, 131
Shah, Mohammad Athar, 172
Shahid, Ahmed, 141, 175
Shannon, Michael, 154
Shao, Jun-Been, 129
Shao, Tzu-Fann, 110
Shapiro, Julie R., 160
Shapiro, Sidney, 34, 60, 107
Sharma, Anup Kamal, 174
Sharp, James Clayton, 161
Shastry, Usha, 172
Shearing, Tracy Leigh, 163
Shen, Zhi-Jian, 177
Shepherd, Kenneth G., 28, 127, 146
Shepherd, Walter E., 128
Sherer, Adam Dale, 137
Sherman, Rita, 139
Shern, Yihchih, 170
Shi, Larry, 97
Shields, David Edward, 76, 151
Shigihara, Takashi, 131
Shipkowski, James Paul, 132
Shore, Jonathan Paul, 155
Shostak, Murray, 128
Shreeniwas, Radha, 175
Shulman, Kenneth Arthur, 132
Sidare, Matthew, 178
Sidas, Zita J., 172
Siegel, Kevin Paul, 165
Siegel, Marc D., 161
Siegel, Stephen Jon, 163
Sierra, Jose Manuel, 131
Silkman, Ronald Wayne, 140
Silver, Matthew Howard, 173
Simcox, Donald G., 28, 146
Simhi, Menashe, 28, 126
Simon, Henry I., 147
Simonds, William James, 165
Simonian, Dmitri, 141
Simonson, John, 96
Simpson, Kevin A., 172
Simpson, Mario T., 178
Sin, Eng Joo, 156
Sinai, Julian, 136
Singh, Daljeet, 130
Singh, Harvinderpal, 172
Singh, Randolph N., 178
Singh, Saurabh Pratap, 176
Sinko, James William, 110
Sirgany, Wadie N., 149
Siu, Eddie Kam Wah, 160
Skola, Timothy L., 148
Skovorada, Andre, 76
Slawson, Martin Kenneth, 138
Slaymaker, Frank, 56
Slocum, Janet E., 156
Small, Jeffrey A., 123
Smith, Andrew L., 160
Smith, David Russell, 131
Smith, Jason Stephen, 142
Smith, Mark Wesley, 132
Smith, Mary A., 152
Smyntek, Kevin Paul, 163
Snider, Raymond S., 41
Snyder, John Somerville, Jr, 131
So, Hing-Cheong, 25
Sobel, Elliot, 158
Sobolewska, Bozenna, 98
Sobolewski, Roman, 64
Sobski, Andrzej, 139, 170
Soderman, Donald A., 148
Soderman, Richard Jennings, 150
Soerensen, Ole Hoffmann, 56
Soh, Andrew T.H., 158
Solis, Luis Alejandro, 140
Song, Jingqing, 141
Song, Yanhai, 143
SooHoo, Spencer L., 150
Sorensen, Robert Henry, 158
Sosinski, Gregory C., 134
Soulos, Thomas Steven, 133
Souriau, Laurent, 98
Souza, Kip Anthony, 150
Soyata, Tolga, 140
Spencer, Marc Douglas, 175

Spencer, R. Mark, 152
Sperber, Gerald Leslie, 128
Sperber, Michael A., 141
Sperry, Robert Hammond, 116, 133
Speth, William M., Jr, 133
Spiegel, Richard J., 152
Spinoza, 43
Spivey, Chubasco C., 176
Sproull, Robert L., 46
Sreepada, Madhu Sudhana Rao, 130
Srivastava, Pranjal, 176
Sroka, David, 159
Staelin, Carl Hudson, 163
Stairman, Edward Mark, 165
Stancampiano, Charles V., 54, 113, 130
Standera, William James, 150
Stankus, Alan Joseph, 128
Stanojevich, Bob Srbislav, 119, 140
Stanton, Kevin Paul, 171
Stark, Lawrence, 30
Stark, Richard, 56
Stasko, Joseph, 129
Staudenmayer, Harold F., 28, 146
Steckl, Andrew J., 130
Steen, Robert F., 147
Steinberg, Kenneth Ivan, 170
Steinthal, Michael Gregory, 165
Stephens, Daniel E., 177
Stephenson, F. William, 55
Stephenson, Kendall Alvin, 97, 119, 137
Stephenson, Robert Ragan, 165
Sternbane, Laurence Jon, 163
Sterritt, Janet, 155
Stevens, Kenneth C., 153
Stevenson, F. William, 56
Stevenson, Robert Lewis, 101
Stoller, Mark Seth, 160
Stolwijk, Jan, 30
Stone, Jeffrey Scott, 159
Stonner, Andrew Weldon, 172
Strassner, John Thomas, 178
Straub, Francis Joseph, 153
Strauss, Burton Mahler, III, 158
Streb, Charles Alfred, IV, 173
Streifer, William, 25
Striemer, Christopher Carl, 143
Stroh, W. Richard, 24
Strongwater, Allan, 152
Stroweis, Jacques, 134
Struebel, Mark Andrew, 157
Stueber, Lawrence John, Jr, 153
Stuhr, Robert Laurence, 154
Sturtz, Robert Joseph, 172
Sud, Ashok, 130
Sudol, Ronald Joseph, 130
Sulaiman, Edwin, 177
Sulatycke, Peter Daniel, 166
Sullivan, Gregory Scott, 178
Sullivan, John Edward, Jr, 148
Sumino, Yoichi, 138
Summers, Christopher Wayne, 121, 138
Summers, James E., 29, 147
Sun, Din-Yi, 178
Sun, Zhaohui, 142
Sungurtekin, Ali Ugur, 134
Supnithi, Pornchai, 175
Surowiec, Charles Matthew, 141
Surya, Charles, 115, 134
Susch, Nicholas Leon, 172
Suthar, Shamal L., 131
Swarthmore, 69
Swartout, George W., 133
Swartz, Alan Jay, 157
Sweetman, Richard Edward, Jr, 161
Swenson, Erick Noak, 18, 146
Swisher, Scott, 31
Swoyer, Vincent, 45
Sypniewski, Jozef, 77
Szabo, Thomas Laszio, 129
Szczerbinski, Timothy, 158
Szeto, Johnny Ka-Lun, 169
Szeto, Theresa Poy-Hing, 161
Szolyga, Thomas Herman, 151

Tabei, Makoto, 96
Tague, Lyle M., 133
Tahir, Abbas, 177
Tahir, Syed Mohammad Ali, 178
Takahashi, Gennousuke, 177
Takahashi, Makoto, 139
Takita, Mark Kenji, 165
Taku, Toshiharu, 38

Talbot, Samuel, 30
Talley, Timothy Jonathan, 168
Tam, Raphael Kwok-Kiu, 153
Tamez-Peña, José Gerardo, 123, 142
Tan, Chak L., 121, 138
Tang, Chi-Kin, 177
Tang, Kit-Ming Wendy, 117, 136, 165
Tang, Tianwen, 142
Tang, Xiangyang, 143
Tang, Xiaoou, 138
Tanverdi, Cengiz, 150
Taradash, Matthew Lawrence, 163
Tausanovitch, Nicolas, 163
Taves, John B., 163
Tawil, Victor, 131
Taylor, Cordigon Cornel, 174
Taylor, James A., 147
Taylor, James H., 29, 127
Taylor, Lawrence Steven, 143
Taylor, Robert B., 18, 146
Taylor, William John, 129
Teitsch, Scott David, 179
Tekalp, A. Murat, 73
Teowee, Gimtong, 163
Terzioglu, Atilla, 165
Terzioglu, Esin, 173
Testerman, Roy Lee, 110, 127
Teumer, Christopher Robert, 164
Thanawala, Akshat Ajay, 177
Thompson, Brian J., 52, 100
Thompson, George Lemar, Jr, 127
Thompson, Horace, 77
Thompson, John Robert, 133
Thompson, John Stewart, 110, 127, 129
Thompson, Raymond L., 16
Thompson, Whitfield Robinson, 141, 172
Thomsen, Steven P., 142
Thorne, Richard Edmund, 161
Thurstone, Frederick L., 77
Tiede, John F., 147
Tilove, Robert Bruce, 114, 132, 152
Tindall, Gerald, 69
Tirone, Thomas Stephen, 166
Titlebaum, Edward L., 34
Tobias, David Aron, 165
Todd, Mark Edward, 164
Todd, Robert A., II, 171
Toklu, Candemir, 122, 140
Tombs, Thomas Nathaniel, 118, 137, 166
Torng, H.C., 47
Torre-Sanchez, Fernando, 135
Torres, Alexandra, 179
Tourassis, Vassilios D., 73
Tousley, Bradford Clark, 119, 138
Towner, Paul D., III, 148
Towsner, Robert Steven, 148
Trainor, Paul Vincent, 150
Tran, Thuy Thi, 153
Treesh, Frederick Hoff, 165
Trietley, Harry L., Jr, 148
Trinh, Huynh Cao, 153
Troise, Richard Frederick, 150
Trott, Marvin, 18, 146
Trotter, John Richard, 157
Truax, Craig Carlton, 173
Tsai, Tzong-Maw, 171
Tsang, Hung, 173
Tse, Jackal Jeun-Cheung, 155
Tsoukkas, Apostolos, 166
Tsumura, Norimichi, 96
Tsybeskov, Leonid V., 96
Tuan, Hsing Chen, 130
Turner, Allyn J., 179
Tushman, Jonathan David, 177
Tuthill, Theresa Ann, 117, 135, 161
Ty, Jonathan S., 171
Tyree, Theodore Kendrew, 172
Tzau, Raymond Chi Sin, 131, 152
Tzau, Tierry, 157

Ucer, Kamil Burak, 122
Ucok, Hikmet Emre, 175
Udy, Karen, 169
Urfer, Kenneth D., 127
Urino, Yutaka, 98

Vakil, Nimish, 69
Valentine, Alan, 15
Van Beek, Petrus, 98
Van Leuvan, Christopher Harold, 175

Van Voorhis, Harold R., 147
Vanacker, Torney Mark, 153
VandenBrande, Jan Hendrik, 117, 134
Vandeusen, Mark Pullen, 154
Vandyshev, Jury, 97
Vanheel, Nicholas Martin, 161
VanNostrand, S. Lance, 138
Vanremmen, William John, 174, 177
Vargas, Urso Atila, 179
Vellani, Amirali, 171
Venable, James Paul, 149
Venetsanakos, James, 174
Verdico, Frank Michael, 179
Verner, Bruce David, 165
Vernik, Igor, 98
Veruki, Margaret Lin, 162
Verwey, James R., 29, 126
Viahos, James Theodore, 177
Viau, Paul Emile, 150
Vieira, Robert Alan, 158
Vipler, Howard, 149
Vlietstra, Justin Eric, 178
Voelcker, Herbert B., Jr, 25
Vogel, Benjamin Moses, 159
Vogel, Howard J., 151
Volz, Phillip Anthony, 171
Von Bahren, Jens, 99
Von Foerster, Heinz, 30
Von Gierke, Henning, 31
Vona, Daniel Francis, 174
Vosteen, Robert Earle, 17
Vukovic, Nada, 123, 140
Vystavkin, Alexander, 51

Waag, Robert C., 34, 107
Wafler, Walter F., 138, 166
Wagmeister, Lee Samuel, 168
Wahba, Brent Jack, 168
Waldrop, Jeffrey Alan, 173
Wall, Thomas Bergan, 160
Wallis, Robert Hall, 149
Wallis, Wilson Allen, 20
Walsh, William Stanford, 129
Walter, Thomas J., 110, 127
Walters, David James, 169
Wan, Angie Mei-Ching, 137
Wan, Lisa Mei-Sheung, 165
Wanamaker, Robert Joseph, Jr, 158
Wang, Chao, 179
Wang, Chia-Chi, 121
Wang, Chung-Chian, 127
Wang, Jing, 170
Wang, Michael, Jr, 157
Wang, Muge, 143
Wang, Rachel Xiaoyang, 140
Wang, Xiaohui, 121
Ward, Denham S., 97
Ward, Joseph, 137
Waring, Bruce Robert, 168
Wariyapola, Pubudu Chaminda, 141
Warker, David Wayne, 155
Warmuth, Matthew William, 172
Warner, H., 30
Warnke, David Lyndon, 173
Warwick, Alan Marshal, 166
Washburn, Neil David, 168
Washizu, Masao, 97
Waterman, Todd Andrew, 165
Watkins, Donald Leigh, 153
Watson, Bruce Allen, 153
Watt, Gordon James, 13
Watts, Edgar Henry, Jr, 148
Waugh, John B., 41
Waugh, Richard, 93
Weaver, Kenneth James, 174
Webster, John G., 110, 127
Webster, Marc Warren, 165
Weeras, F.B. Amitha Buddhika De Silva, 178
Wegman, Charles George, 153
Wei, Kehui, 134
Weibel, Marjorie Eleanor, 166
Weidman, James, 164
Weikel, Donald Josiah, Jr, 148
Weiler, Judith Ann, 165
Weimer, David Leo, 151
Weinberg, Robin Alicia, 175
Weiner, David Howard, 156
Weiner, Mark Steven, 168
Weinman, Vivyan Lee, 169
Weisberg, Jeffrey Alan, 140
Weiss, Marc Steven, 37, 111, 127
Welch, Daniel A., 168
Welder, Tammy Sweeney, 169

Wells, Peter N.T., 56
Weng, Xiaozhen, 122, 140
Wengler, Michael, 64
Weppner, Matthew Brennan, 164
Werner, John Edward, 169
Wesley, David R., 147
Wesolowski, Carl R., 160
Westkirk, Angus Williams, 148
Wheeler, Warren Ray, 14
Wheeless, Leon L., Jr, 29, 56, 109, 126
Whitaker, John Firman, 116, 134
White, David B., 152
White, Donald Ernest, 135
White, Donna Marie, 169
White, Walter W., 29, 126
Whiting, Carlyle F., 40
Wich, Grosvenor S., 18, 145
Wickenheiser, Jeffrey Victor, 162
Wiepert, Mathieu, 166
Wilcox, Jeffrey Randall, 176
Wilcox, Jeffrey William, 172
Wilcox, Robert Harvey, Jr, 157
Wild, Charles Richard, 165
Wilhelm, Neil C., 54
Williams, Carlo Anthony, 141, 174
Williams, Douglas H., 127
Williams, Roy, 71
Williams, Stephen Gareth, 169
Willis, P., 30
Willnecker, Brian Paul, 174
Wilson, David, 31
Wilson, Joseph C., 20
Wilson, Marshall Crawford, 172
Wiltse, Mark Donald, 166
Winter, Christopher G., 154
Wirsig, Stanley Smith, 155
Wolf, Emil, 41
Wolfeld, Warren Scott, 157
Wolff, Christopher Morris, 164
Wong, Linda, 173
Wood, James Albert, Jr, 12
Wood, Joseph Peter, 156
Woods, William S., 159
Woolever, Gerald Francis, 129
Worth, Joseph H., 127
Worthington, Mark William, 157
Wortman, John M., 172
Wronski, Leszek Dariusz, 138
Wu, Andy Tin-Ho, 142, 177
Wu, Chuan-Hui, 138
Wu, De-xin, 77
Wu, Di, 141, 174
Wu, Jian, 122, 139
Wu, Jianbin, 122, 141
Wu, Ming Jian, 174
Wu, Peter Yick Fai, 156
Wyner, Donald Stuart, 149

Xia, Minghui, 143
Xie, Xiandong, 119, 139
Xiong, Wei, 121, 139
Xu, Bingxiong, 143
Xu, Songtao, 143
Xu, Yaowu, 143

Yakop, Masri Bin, 172
Yang, Chau-Jye, 177
Yao, Meng, 121, 138
Yao, Rong, 142
Yarkoni, Barry N., 152
Yates, Arthur Gould, 89
Yeaple, Ronald N., 76, 115
Yeung, Fai, 122, 141, 174
Yin, Gu Feng, 178
Yiu, Kai-Ping, 130
Yokoyama, Ryuzo, 111
Young-Sheppard, Iris, 155
Young, Daniel, 99
Youngs, Kurt William, 162
Yu, Chen Kuang, 171
Yu, Paul, 30
Yu, Qing, 122, 142
Yuan, Frances H., 173

Zacharias, Margit, 96
Zacher, David M., 172
Zachmann, Peter Harold, 28, 125
Zadarlik, Peter P., 152
Zahorian, Stephen Andrew, 149
Zak, Stuart Curtis, 169
Zalonis, Martin Ross, 129
Zandman, Jerald, 147

Zaremba, Andrew J., 152
Zaucha, David Edward, 131
Zeck, Normal W., 135
Zeevi, Yehoshua Y., 128
Zell, Thomas Baylis, 138
Zeng, Wen-Sheng, 97
Zeppos, Cheila, 152
Zhang, Daofa, 141
Zhang, Ren-Yu, 144, 178
Zhang, Zhao-Nan, 118
Zheng, Menghui, 62, 140
Zheng, Zhe, 105, 176
Zhou, Xing, 117, 136
Zhou, Xingxiang, 143
Zhu, Qingyuan, 143
Zhu, Wei, 142, 177
Zhuang, Ning, 143
Ziegler, Eric Martin, 167
Zigadio, Joseph P., 129
Ziobro, James M., 138
Zuber, Martin James, 158
Zucker, Robert Michael, 143
Zuckerman, Paul R., 56
Zugger, Michael Eugene, 162
Zweig, Kenneth Edmund, 167
Zwieslocki, James, 31
Zwolinski, Stephanie Marie, 155
Zwolinski, William Jeffrey, 159